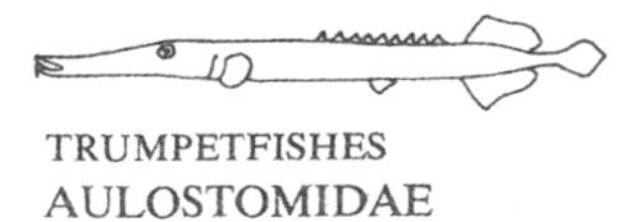

TRUMPETFISHES
AULOSTOMIDAE

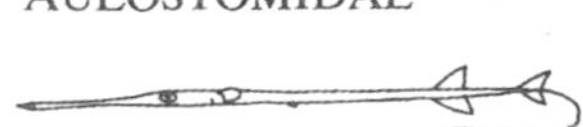

CORNETFISHES
FISTULARIIDAE

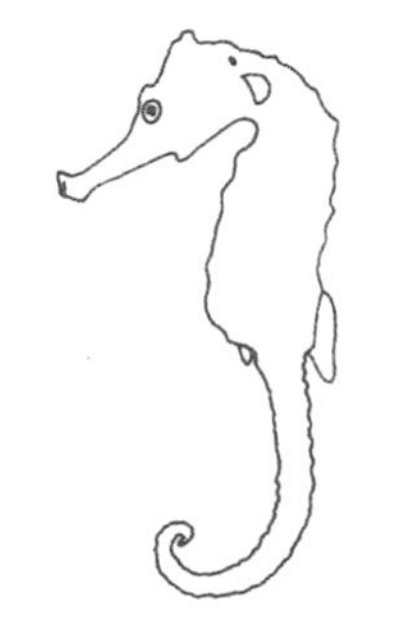

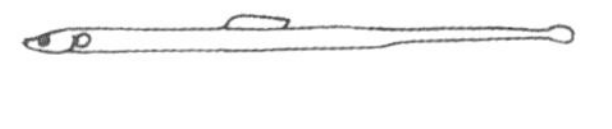

SEAHORSES & PIPEFISHES
SYNGNATHIDAE

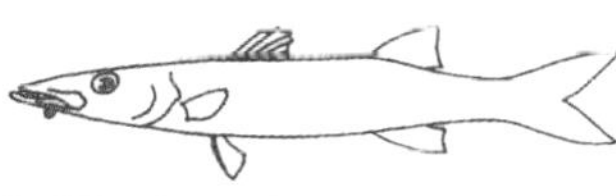

BARRACUDAS
SPHYRAENIDAE

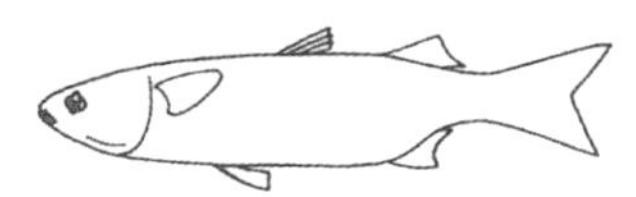

MULLETS
MUGILIDAE

THREADFINS
POLYNEMIDAE

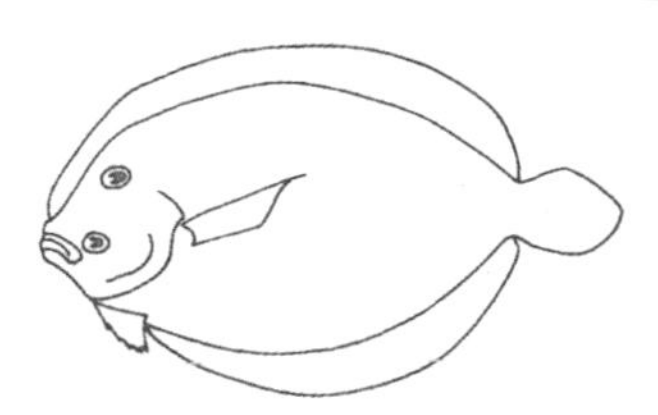

LEFTEYE FLOUNDERS
BOTHIDAE

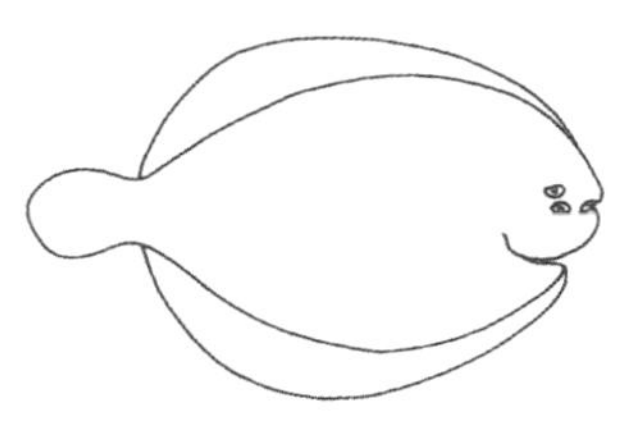

SOLES
SOLEIDAE

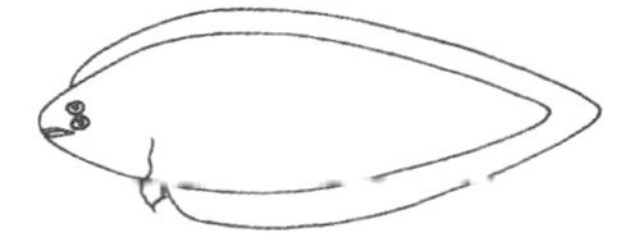

TONGUEFISHES
CYNOGLOSSIDAE

CARDINALFISHES
APOGONIDAE

SNOOKS OR ROBALOS
CENTROPOMIDAE

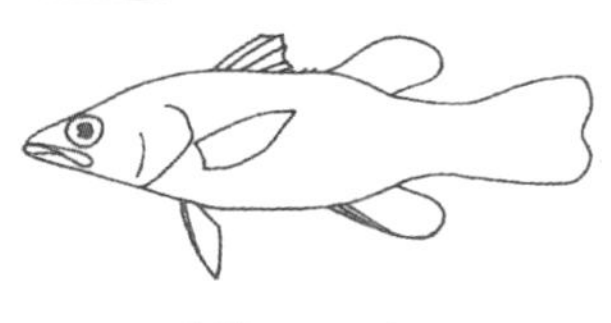

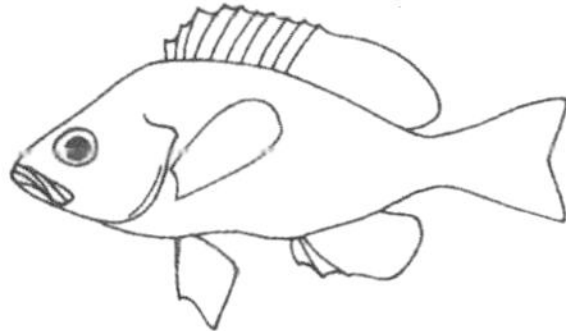

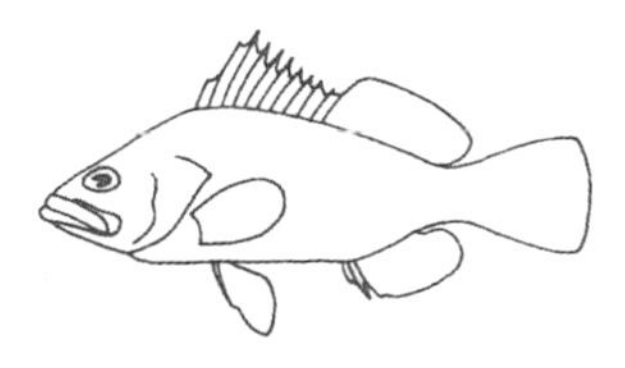

SEA BASSES
SERRANIDAE

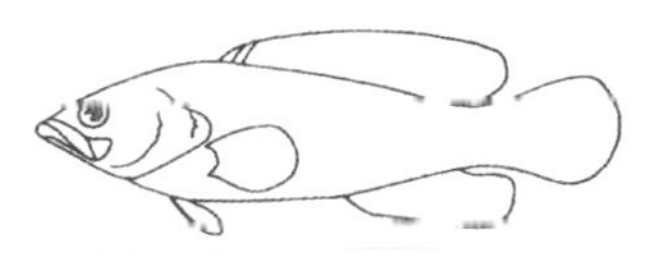

SOAPFISHES & ALLIES
GRAMMISTIDAE

BASSLETS
GRAMMIDAE

FULL *fathom* Five

Full fathom five thy father lies;

Of his bones are coral made;

Those are pearls that were his eyes;

Nothing of him that doth fade,

But doth suffer a sea-change

Into something rich and strange.

Sea-nymphs hourly ring his knell:

Ding-dong.

Hark! Now I hear them—Ding-dong, bell.

—William Shakespeare,
"Ariel's Song," *The Tempest*

Also by Gordon Chaplin

Dark Wind: A Survivor's Tale of Love and Loss

Fever Coast Log: At Sea in Central America

Joyride

FULL

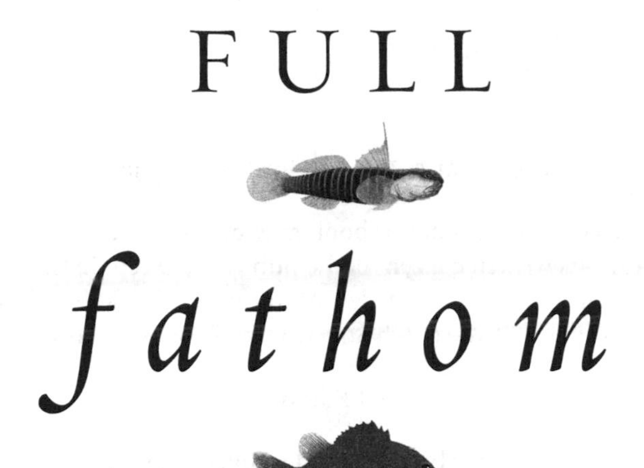

fathom

Five

Ocean Warming and a Father's Legacy

GORDON
CHAPLIN

Arcade Publishing • New York

First Edition

Arcade Publishing books may be purchased in bulk at special discounts for sales promotion, corporate gifts, fund-raising, or educational purposes. Special editions can also be created to specifications. For details, contact the Special Sales Department, Arcade Publishing, 307 West 36th Street, 11th Floor, New York, NY 10018 or arcade@skyhorsepublishing.com.

Arcade Publishing® is a registered trademark of Skyhorse Publishing, Inc.®, a Delaware corporation.

Visit our website at www.arcadepub.com.
Visit the author's website at www.gordonchaplin.com

10 9 8 7 6 5 4 3 2 1

Library of Congress Cataloging-in-Publication Data

Chaplin, Gordon.
Full fathom five : ocean warming and a father's legacy / Gordon Chaplin.
pages cm
ISBN 978-1-61145-895-4 (alkaline paper)
1. Bahamas—Environmental conditions. 2. Global warming—Bahamas. 3. Ocean temperature—Bahamas.
4. Marine fishes—Ecology—Bahamas. 5. Böhlke, James E. Fishes of the Bahamas and adjacent tropical waters. 6. Chaplin, Gordon. 7. Chaplin, Gordon—Family. 8. Chaplin, Charles C. G., 1906–1991—Influence. 9. Fathers and sons.
10. Scientists—Bahamas—Biography. I. Title.
GE160.B2C42 2013
551.46092—dc23
[B] 2013022518

Printed in the United States of America

I would like to dedicate this labor of love, admiration,
and growing ecological awareness to my father,
Charles C. G. Chaplin, who is its hero.

Also to his colleague, friend, and mentor,
James E. Böhlke, who gave his life for science.

Contents

PART 2

PART 3

DRAMATIS PERSONAE

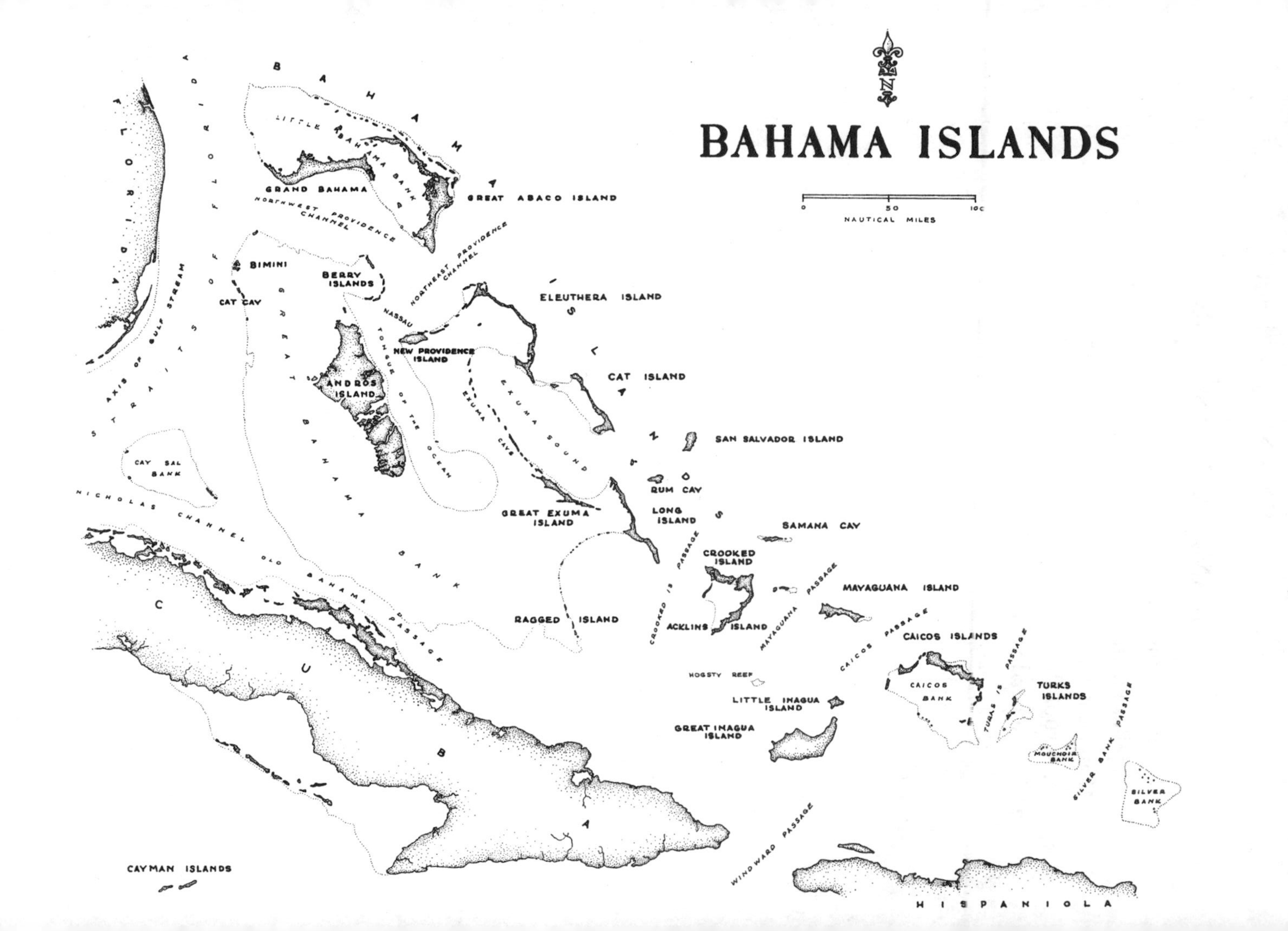
BAHAMA ISLANDS
0
50
10C
NAUTICAL MILES
FLORIDA
STRAITS OF FLORIDA
AXIS OF GULF STREAM
BAHAMA ISLANDS
LITTLE BAHAMA BANK
GRAND BAHAMA
NORTHWEST PROVIDENCE CHANNEL
GREAT ABACO ISLAND
NORTHEAST PROVIDENCE CHANNEL
BIMINI
CAT CAY
BERRY ISLANDS
NASSAU
NEW PROVIDENCE ISLAND
ELEUTHERA ISLAND
CAT ISLAND
ANDROS ISLAND
TONGUE OF THE OCEAN
EXUMA CAYS
EXUMA SOUND
SAN SALVADOR ISLAND
RUM CAY
LONG ISLAND
GREAT EXUMA ISLAND
SAMANA CAY
CAY SAL BANK
GREAT BAHAMA BANK
NICHOLAS CHANNEL
OLD BAHAMA PASSAGE
CROOKED ISLAND
CROOKED IS PASSAGE
MAYAGUANA PASSAGE
MAYAGUANA ISLAND
RAGGED ISLAND
ACKLINS ISLAND
CAICOS PASSAGE
CAICOS ISLANDS
CAICOS BANK
TURKS IS PASSAGE
TURKS ISLANDS
HOGSTY REEF
LITTLE INAGUA ISLAND
GREAT INAGUA ISLAND
MOUCHOIR BANK
SILVER BANK PASSAGE
SILVER BANK
CUBA
WINDWARD PASSAGE
CAYMAN ISLANDS
HISPANIOLA

Acknowledgments

There are many people without whose help this book never could have been written. First and foremost is Dr. Katriina Ilves, who, during her stint as a Chaplin Fellow at the Academy of Natural Sciences of Philadelphia, whipped all the science into coherent shape and authored the two ground-breaking papers I've liberally quoted from. Many drafts of this book were uncomplainingly read by my wife, Sarah Teale, my friend and advisor, Michele Slung, and my tireless agent, Meg Thompson. Also weighing in with invaluable advice and support were Thomas Powers, Tom and Mary Edsall, Alan Rinzler, and Ed Wells. At the Academy of Natural Sciences of Philadelphia, I owe a huge debt to Dr. John Lundberg, the curator of fishes, who not only came up with the idea but helped keep it alive at critical points. Expedition scientists Dominique Didier Dagit, Heidi Hertler, Danielle Kreeger, Loren Kellogg, Walter Jaap, Mark Westneat, and Ron Eytan made up the front lines. And in the Bahamas we depended on the good will, trust, and generosity of Michael Braynen, director of Fisheries, Casuarina McKinney, executive director of the Bahamas Reef Environment Educational Foundation, the late Sir Nicholas Nuttall, its founder, Dr. Kathleen Sullivan Sealey, and her husband, Neal. My sister, Susan, the only person in the world who shares many of these memories with me, was there physically and spiritually to keep me on track. Thanks to all from the bottom of my heart.

The text and images for the yellowmouth jawfish, the greenbanded goby, the fairy basslet, the blackcapped basslet, the cherubfish, and the papillose blenny, from the book *Fishes of the Bahamas and Adjacent Tropical Waters*, by James E. Böhlke and Charles C. G. Chaplin, copyright © 1968 by the Academy of Natural Sciences of Philadelphia, and the photograph of James E. Bölhlke on the Amazon River, are reprinted with permission of the Academy of Natural Sciences of Philadelphia.

Excerpt from "Two Tramps in Mud Time" from the book *The Poetry of Robert Frost*, edited by Edward Connery Lathem, copyright 1969 by Henry Holt and Company, copyright © 1936 by Robert Frost, copyright © 1964 by Leslie Frost Ballatine, is reprinted by permission of Henry Holt and Company LLC.

The article "Did Fatal Dive Pair Break Their No-Help Pact in Record Bid?" is reprinted by the kind permission of the Nassau *Tribune*.

The map of the Bahama Islands is reprinted with the permission of Scribner Publishing Group from *Bahama Islands: A Boatman's Guide to the Land and the Water* by J. Linton Rigg, revised by Harry Kline. Copyright © 1949, 1951, and 1959 by D. Van Nostrand Company, Inc. Copyright © 1973 by Charles Scribner's Sons. All rights reserved.

PART I

An Unexpected Turn

The Call

August 2003

FIFTY YEARS AGO, WHEN I was a child, I lived on an island in the Bahamas, in an old wooden bungalow with wide verandahs that fronted on the sea. At one end of the curving strip of white sand that separated our house from the ocean was an ancient white lighthouse marking the entrance to Nassau Harbor. At the other fluttered the parasols of a little concession called Paradise Beach, where water taxis took day-trippers from town.

The limestone island, originally a coral reef, was low and scrubby, but many generations of settlers had planted it with palms, sapodilla, yellow elder, poincianas, hibiscus, guava, banyans, frangipani, casuarina, gumbo-limbo, jasmine, bougainvillea, agave, and oleander. The thick, salty air of the easterly trades that rattled in the palms smelled of flowers on the verge of decay.

The ocean pushed in on this beach with darkening shades of color: aquamarine giving way to emerald, finally changing to sapphire. On sunny days, the colors glowed as if backlit from below. In winter the shallow bar half a mile offshore would produce dark blue storm breakers that reared up like horses, white manes blowing behind them. The ocean was the defining element of this place: its moods were your moods. It was not a drab, oppressive, monolithic element. Just the opposite. It went to hallucinogenic extremes, like the Beatles' *White Album*.

Through the clear water, you could sometimes see the dark shapes of large fish—barracudas, rays, the occasional shark—casting their shadows on the sandy bottom. Sometimes a cloud of blue fry minnows would scatter itself across the surface to escape the swirl of jacks below. If there was surf, sometimes you could see the black dorsal fins of silvery gafftopsail pompanos outlined in a transparent wave.

The water temperature never dropped out of the seventies, and in the summer rose into the eighties, just a little cooler than body temperature. In the water you could do things that you only do in your dreams, and entering it made you feel as if you were coming home. *Here I am, where I ought to be.*

Like many children of the sixties, I've always had trouble with the idea of home. As a young man, it represented everything I was trying to leave behind: the old values, the old traditions, the old legacies. I'd taken Kerouac to heart, and I wanted the open road.

You never completely recover from that wild exhilarating ride. Later, when you want to put down a root or two, no place feels exactly right. If you have a family and are not so mobile anymore, you might find yourself stuck. You try to make the best of it, but you know your real home is somewhere else. If it exists at all.

To complicate things even more, I loved the place where I grew up. I have an uneasy feeling that the home I'm looking for now will never measure up to it, and that the legacy that was passed to me there will slowly fade away.

In this New York loft on the fringes of Soho, where I've washed up after a shipwreck, there are quite a few things I inherited from my father. They seem lonely in the rough open space with its bare brick walls, as out of place as I feel myself. A hand-carved model of the old

Bristol Channel cutter that he sailed from England in 1936 is taking up floor footage. In a mahogany vitrine propped against the wall, a collection of cone shells from all over the world. Tacked into the bricks above them, a three-by-four- foot oil painting of a barracuda in deep blue water.

Stuffed in various nooks and crannies are two rosewood Japanese carp, an ivory crawfish, scrimshawed whales' teeth and walrus tusks, and a variety of articulated silver and brass fishes and whales. By the door, a narwhal horn hatstand. Under the back windows, a five-foot mosaic of another barracuda set in a three-hundred-pound slab of cement.

And in the bookshelf, a massive leather-bound collector's edition of my father's life's work: *Fishes of the Bahamas and Adjacent Tropical Waters.*

If I get up from my desk and walk to the back windows beyond the barracuda mosaic, I can look straight down Delancey Street to the Williamsburg Bridge, with a sliver of salt water running underneath. The sliver is flat gray most of the time, but at dawn it can be red. Sometimes a small freighter will drift under the span, sometimes a sailboat. It's all I can see now of my father's legacy to me . . . a legacy that I first turned my back on and that later betrayed me with a typhoon that took my partner and my boat. Coming to terms with it is seeming less and less likely.

Today is a bona fide New York August dog day, and heat waves rising from the black asphalt of Delancey Street cause the distant sliver of ocean to ripple like a mirage. The old loft building creaks and groans.

Back at my desk, words dance wearily on the screen of my computer. The novel I'm working on isn't coming together, and I'm beginning to fear it never will. When the phone rings, I'm only too glad of the distraction. "Is this a good time to talk?" asks an effervescent female voice.

The voice belongs to Dominique Didier Dagit, associate curator of fishes at the Academy of Natural Sciences in Philadelphia, where my father had worked in the fifties and sixties. I'd met her a few times at academy functions. Fortyish, blond, and vivacious, she seemed about as far from the stereotypical scientist as you could get.

"I'm so excited I can hardly put the words together!" She sounds even less like a scientist than I remember. "You know your father's work here with Jim Böhlke?"

I look at his imposing volume on my bookshelf and tell her I do.

"There's really been nothing like it, before or since. They spent fifteen years collecting, you know. It was a landmark work."

"Yes. I was part of it."

"Of course. Their archives are huge . . . specimens, photographs, films, field notes . . . unique in the field. All that material is here at the academy, perfectly preserved, totally accessible."

"Still there, hunh?" I remember trotting behind my father down the long rows of labeled specimen jars on the dusty, dim shelves.

She trills a laugh. "They'll be here when we're dead and gone, Gordon. So . . . are you ready for this?"

She outlined how her boss, John Lundberg, the head curator, had discovered many old films in the archive and had suggested they might have recorded a baseline against which change could be measured. She had been working with the field notes to pinpoint collecting sites. "Our idea is to use all this data to make a retrospective study of how the reefs have changed since your father's collections. As far as I know, it'll be the first fifty-year study ever made."

"Really?" My hand around the phone suddenly cramps. "It does sound exciting. And where do I come in?" She's probably looking for funding.

"You are the key, Gordon," she says. "The last surviving member of the collecting team. We aim to go back to the original sites to make our own collections, and you are the only person alive who knows exactly where they were. You do remember, don't you?"

My mind is already underwater. Holding its breath. "As well as I remember the rooms in our old house."

I don't ask about funding. Instead, I make a date with Dominique to go down to Philadelphia, meet her two colleagues for this project, and discuss how things might work out.

To calm down a little and think all this over from a broader perspective, I climb the apartment building's stairs to the roof (I watched the World Trade Center towers fall from here). In the shade of an old wooden water tower, looking across the rooftops to the golden female figure on the spire of the Manhattan Municipal Building, it occurs to me that my life might be about to take an unexpected turn.

In many ways, my father's original study is my sibling. I spent my childhood with it in the Bahamas, went on many of the collecting expeditions, watched it take shape and grow alongside me, and remember my first foray below the surface with almost numinous clarity. In brand-new goggles, flying down to where my father beckoned from below, I found myself in one of those life-changing moments when the world rearranges itself into new priorities.

If I understand Dominique correctly, I am being offered the chance to pass back through the looking glass into my childhood, with scientists carrying the bags. As the sole surviving member of the original team, I'd be able not only to guide the scientists to the old collection sites but also to help them recreate our methods, make informal fish density and coral cover comparisons, and fill in historic cultural detail. Take up where my father left off. Pay my debts. Keep his legacy alive.

But I haven't been back in thirty years, not since being banned from the Bahamas for life for bringing in a quarter ounce of marijuana (though the ban eventually was lifted), and the prospect of returning if only for a series of short expeditions is daunting on many levels.

As the man said, you can't go home again. What if my sacred memories turn out to be false or faulty? What if I myself have changed too much to fit back in? And scariest of all, what if the place I'm returning to no longer exists?

But I've been haunted by this place all my life. At the very least, I might be able to put some ghosts to rest.

Under the Radar

September 2003

IT'S A TWO-HOUR DRIVE FROM downtown New York to Philadelphia. A little past ten a.m. on a crackling early fall Monday I meet Dominique in the somber basement Chaplin Library of the Academy of Natural Sciences' ichthyology department. Bookshelves full of reference works and academy proceedings line the walls, and papers are stacked high on a large conference table.

Near the door, a brass plaque proclaims the library was donated by my mother, who also endowed the Chaplin Chair for the department's curator and underwrote the cost of publishing *Fishes of the Bahamas.* The art on the walls is eerily familiar: an illustration by artist Steven P. Gigliotti from the text of *Fishes*, of an elongated spotted brown-and-white goby named *Acanthemblemaria chaplini* in honor of my mother, and a painting of two large iguanas collected in Cuba in 1948 on an expedition funded by my uncle Cummins Catherwood. We'd kept one of these iguanas in a cage in our house for a few weeks, feeding it bits of chicken, but it stayed fierce and wild and finally we gave it to the zoo.

Dominique, dressed informally in tan slacks and a light blue sleeveless blouse, gestures to the two-foot-high pile of notebooks on the library table: "Those are the Böhlke-Chaplin field notes. Every collection station they ever made. Take a look."

It's the first time I've seen them, all in my father's or Jim Böhlke's neat (almost identical) handwriting, and often including little sketches of locations. They number the sites (724), enumerate collecting conditions and techniques, list personnel, and often provide additional comments at some length.

It begins to feel as if my father and Jim Böhlke are in the Chaplin Library with us, along with the brass plaque and the art on the walls. All a bit overwhelming.

As I'm poring over the notebooks, noticing that my sister Susie and I are often listed among the collecting personnel, two more women walk into the library. Dominique introduces them as Heidi Hertler and Danielle Kreeger from the academy's applied ecology department, the Patrick Center.

The three scientists turn out to be an unconventional, accomplished trio. Dominique, forty, is an authority on a rare deepwater shark relative called the ratfish; Heidi, thirty-nine, studies the impact of land use on marine environments; and Danielle, forty-three, researches aquatic ecosystems. All of them have doctorates.

But one would never take any of them for scientists. Dominique has the high-powered effervescence of a fashion executive; the dark-eyed, breathy Danielle has recently completed her final treatments as a man-to-woman transsexual. Heidi lives on a boat in the Delaware River with her boyfriend, Clarence, an African American rugby champion, and is built like a rugby champion herself.

The science involved in this survey, as the three describe it, would be pretty straightforward. We'd return to a series of locations where reliable measurements have been taken in the past, take new comparable measurements, and then study the ecological changes. John Lundberg's idea to use the archival films could give an added dimension, depending on quality and subject matter.

Ecology, the study of dynamic environmental relationships, was in its infancy in my father's day. We were simply trying to collect and describe species of reef fish: all we could possibly find. How these fish fitted into a community, what that community was, and how it might be changing wasn't really our concern. The concept of biodiversity—that ecological health depends on the survival of as many living organisms as possible—was something we instinctively agreed with, but did not formally understand.

So the work at hand seemed the logical evolution of my father's. It would build on his platform, broaden it, and take it in significant new directions.

Danielle has already compiled an exhaustive spreadsheet of all the historical collection sites. As an associate curator, Dominique has access to the academy's collection of fish specimens assembled by my father and Jim Böhlke, databased phylogenetically by collection site, species abundance, et cetera. Heidi is a world-class field researcher. I am the sole surviving member of the original team. And the fact that we have a fifty-year spread for our study makes it possibly unique. Most other studies, if not all, have to stretch for a thirty-year spread.

However, the devil is in the details. All of Böhlke-Chaplin's significant historical collections were made using an organic fish poison called rotenone. In fact, my father was one of the first to combine the use of rotenone with the use of SCUBA gear, which had just been invented.

In my father's day, a permit for the use of rotenone was not required in the Bahamas. Now it is, and some of her colleagues have advised Dominique that it will be very hard, if not impossible, to obtain. Poisoning fish is not politically correct, number one, and number two, some of the reefs where my father collected are now tourist attractions. Number three, the Bahamas have become quite nationalistic and no longer welcome foreign scientists with open arms. But without the use of rotenone, it would be impossible to replicate my father's collections.

"What makes you think we'll be able to get it?" I ask.

"Well, we won't know until we try." Dominique treats me with her most winsome smile. "At least we'll get a few days on the beach and a good tan."

"Right . . ."

"Seriously, I think your father's name still carries a lot of weight down there."

"Why do you think that?" I'm remembering my father's futile efforts to get my ban lifted thirty years ago.

"I was there on a recent project in Bimini and met some people from the Fisheries Department. They all knew his book. If his son were to explain how we aim to carry on his work, they'd listen very carefully. Of course, we'll make all the data we get available to them. It could be really important data. And as you know, rotenone doesn't harm the reefs and dissipates in a few hours. We'll give them plenty of info on it."

I had made some research calls before this visit to scientific friends of my father's. One question kept coming up, and I pose it to Dominique. "Assuming we do get the permit, what about the variables involved in replicating the historical collections? You know, time of year, ocean conditions, number of collectors, rotenone quality?"

"Yup. The variables." She nods and raises her eyebrows approvingly. "Frankly, some people are skeptical, but we think we can deal with them. John Lundberg agrees."

I've always been in awe of scientific expertise, so my further questioning is not as sharp as it might have been. For example, I never ask how long it might take to complete this study—to make all the collections, do the lab work, and publish a paper. Maybe I'm scared of the answer. After all, my father and Jim Böhlke had taken fifteen years.

"So," I say finally. "The academy is, uh, *behind* the project?"

Founded in 1812, Philadelphia's Academy of Natural Sciences is the country's oldest research museum, the model for the Smithsonian in Washington and the American Museum in New York. Its journal is the first peer-reviewed publication in the US devoted to natural sciences. Its collections include specimens from Cook's voyages and the Lewis and Clark expedition, its library contains reference works dating back to the 1500s. Its ornithology, botany, malacology, vertebrate paleontology, and ichthyology collections rank among the world's most complete. Legendary scientists associated with the academy include Charles Darwin, Thomas Henry Huxley, Thomas Jefferson, John James Audubon, Edward Drinker Cope, Thomas Say, and Ruth Patrick.

When I was a child, the academy was still very much in its salad days, thanks in part to the legendary fund-raising and PR efforts of my parents' friend Dr. Patrick, head of the limnology department, who invented a way of measuring river pollution by examining the resident diatoms.

But in the early nineties, the academy fell on hard times. It had made shaky endowment investments and its city stipend had been cut off. Income from grants, gifts, and exhibits was withering in an apathetic climate. Growing budget deficits sent key scientists and administrators looking for greener pastures. It was forced to sell 18,200 specimens from its historic mineral and gem collection. Its current president, James Baker, is a bureaucrat from NOAA, concerned with lopping off and trimming.

The three women look at each other quickly then look away. "We'll be using the academy's facilities, its labs, equipment, archives, and all that, of course. But we'll be working on our own time. We think this project is worth it," Dominique offers.

"Does Baker know about it?"

"Uh, no. We're kind of under the radar net."

"It's better this way," Danielle says with a grin. "We avoid the bureaucratic hassles. We can pretty much do what we want."

I'm beginning to see the women in a new light: a bunch of renegades, in stark contrast to my father's team, which included not only the academy's director but also the curator of fishes, hired specifically to work with him. Well, that's fine. Under the radar is my style too, and if I'm to carry on my father's work, at least it will be on congenial terms.

Maybe anticipating my next question, Heidi launches into a status report on research funding for museums like the academy, the Smithsonian, and the American Museum of Natural History in New York. The picture is bleak. Under the Bush administration, grant money is drying up, budgets are shrinking, staff is being cut back. The academy budget has been way in the red for many years, and the Patrick Center, which depends almost entirely on grant money, is in the worst shape of all. Heidi herself is about to take a new job with an NGO.

I lean back in my chair and grin at the three renegade women. "So, tell me. What do you think an expedition to the Bahamas would cost, anyway?"

They estimate $10,000 for two weeks for the four of us. "Do you know anyone who might be interested in contributing?" Dominique deadpans. Besides me? Actually I do. There are quite a few family friends and relatives in Philadelphia who were involved in the original collections.

It takes the better part of a year to organize the trip. Fund-raising turns out to be a lot harder than I'd thought. A check for $50 seems the height of generosity for many of those family friends and relatives. I raise about two-thirds of the money, though, and agree to underwrite the rest myself. This would turn out to be just the beginning. Still, even if I'd known then what I know now about the time and trouble in our futures, I would have been just as eager to go.

Welcome Home

June 2004

Fifty years ago, June in Nassau was as close to heaven as I could get. School would be just out, weather hot but not yet sticky, poincianas putting out their festive red flowers all over town. The springtime easterly trades finally would be calming down, the water temperature would have climbed into the eighties, and you could dive for hours without so much as a goose bump.

Now, as my three scientific colleagues and I wait on the dock behind the Club Mango on East Bay Street for a ride across Nassau Harbor to Paradise Island (Hog Island, in the old days, because it was used for livestock), I pick up a powerful whiff of that special June feeling. Everything seems possible, joyous events right around the corner. Summer clouds tower familiarly in the east, and black-and-yellow-striped sergeant major damselfish hang in the water around the pier pilings just the way they used to. My body feels like a teenager's, and my heart pounds with plenty of childlike adrenaline.

For our base of operations, we've been able to rent rooms in the very house where I grew up: an old wooden bungalow on the narrow western end of Paradise Island, now owned by a Bahamian couple I've never met, Ronnie and Joan Carroll, who run it as a bed-and-breakfast. Serendipitously enough, they've kept the name it acquired during our thirty-year tenure there: the Chaplin House.

The stocky, grizzled guy in the half-unbuttoned Hawaiian shirt and frayed khaki shorts piloting a beat-up skiff our way turns out to be Ronnie himself . . . an ex–commercial diver and race-car driver whose family (the Tuckers) has been in the Bahamas since the seventeenth century. So far, so good: I was afraid he'd be a smooth-talking tourist hustler.

Ronnie seems to approve of our project, but he's so gruff it's hard to be sure.

"Nassau's all shot to hell," he offers with a bleak grin, "but we've done our best to keep the old place the way your father left it."

I stand in the bow as we start across the half-mile-wide harbor, still sealed in my June fantasy but beginning to feel the claws of the present frittering away the envelope of the past. Things look more dilapidated and yet more glittery than I remember. The wharves and sheds on the Nassau side are sagging and rusty, docked with tramp freighters from the Caribbean's more notorious fever ports. The famous gin-clear water is now a cloudy green (Ronnie blames this on dredging). Replacing the out-island fishing smacks of my childhood, trading sloops from Haiti with sails of rags drift in on the east wind, carrying, Ronnie speculates, cargoes of illegal immigrants and narcotics.

But, in the center of town, Prince George's dock has been extended to accommodate eleven huge cruise ships at once. And on the Paradise Island side of the harbor, where we're headed, there's a new Club Med four docks east from the Chaplin House. Beyond that, looming above palm trees, casuarinas, and condos, the huge Atlantis resort complex pinkly dominates everything. Almost single-handedly it has dragged Nassau into the twenty-first century and made it the destination of choice for more than five million tourists a year.

The Chaplin House dock has a new gazebo, but otherwise looks exactly the same, in its own eerie time warp. Ronnie's wife, Joan, a Swede and former model whom he met during his turn as a race-car driver, walks out to greet us with a wonderfully genuine smile.

"Welcome home!"

I just nod and try my best to smile back, not trusting myself to speak.

Every step up the long concrete walkway from the dock is a step into the fourth dimension. My first view of the south verandah of the main house is dominated by the untarnished image of my father in his favorite dark blue nylon swim trunks, his tanned back to us, washing snorkeling gear at the tap below the railing and carefully laying it out to dry. He died thirteen years ago at age eighty-four after a ruptured aneurysm. I've brought his ashes with me.

In the cool, familiar shade of the south verandah, Joan hands me a letter my father wrote her after she and Ronnie bought the place in the late seventies. My own hands are shaking so much that I can hardly read it.

> Dear Mrs. Carroll,
>
> Your letter of July 4 arrived yesterday, and we are so delighted and happy to know that our beloved house is now in appreciative hands. It is a marvellous situation, isn't it? Life there was bliss.
>
> As far as I know the house was built by Morton Turtle and belonged to a Mrs. Prince before Mrs. Gardner bought it (we bought it from her daughter in 1948). One of the Agassiz family owned it at one time, for it was known as the Agassiz House. The scientist Alexander Agassiz made a study of Bahamas fishes in 1892 so it is possible he was there before 1910, when he died. The main house is all wood and mostly Abaco pine which is termite resistant and has recently been painted and repaired at great cost.

In addition to putting in the picture window in the living room, I built the aquarium and the substantial dock which I hope is holding . . . unfortunately my caretaker, a frightful little monster named Jerry Adderley, now living in Long Island on the proceeds of robbing me blind, used much of the dock money I sent him for his own purposes and stole my Boston Whaler into the bargain. Honest caretakers have always been a problem in the Bahamas. I'm afraid the house painting is not up to snuff either due to that little worm—may the fire of St. Anthony afflict him in his tenderest parts.

One of the joys of the Chaplin House—do call it that—is the constant change of the sea colors and its calms and roaring surf. Sometimes you get westerly gales—we had one that lasted for 10 days and all the sand was washed away, our outer wall was down to bedrock and we needed a ladder to get down to the beach, but don't get discouraged, the prevailing easterly wind will bring it all back in no time. Anyway it is reassuring to know that the house has been through at least one severe hurricane with the loss of only the dock.

One thing I can heartily recommend is the construction of some artificial reefs at varying distances offshore; I made two, one inshore and one about 20 yards out. All you need is 100 cinderblocks dumped by some bumboat in an irregular pile, in no time at all you will have an interesting little private reef with a very substantial population, watching it grow is fascinating. Do ask for my underwater FISHWATCHERS GUIDE at the local dive shop, you can take it down with you and identify the residents of your reef. Don't let them give you my rival's book by one Jerry Greenberg, it is not nearly so good.

You are probably wondering why we ever left; mostly from skin problems, too much sun is a mistake and my wife has had a lot of trouble and can never go out on the Bahama beaches anymore except in the evenings. My son and daughter in law would have liked the place but as they have recently been divorced it is just as well they didn't have it to quarrel over.

Perhaps you will ask me to come and stay sometime; I can show you all sorts of good places for snorkelling and of course know a lot of people there. Good luck, and I know you will love the place more and more.

Sincerely,
Charles Chaplin

I can't see very well when I finish reading, but after a short silence Joan Carroll's hand touches my arm and I hear her soft voice: "Did he talk like he wrote?"

All I can do is nod.

"I wish I could have known him."

I brush at my cheeks and clear my throat. "He would have loved the way you've kept the house."

But I know he wouldn't have: by his standards the place looks run-down. Unchanged, but run-down. It's exactly the way I myself would have kept it, if I'd been able to take it. (It was my lifetime ban that botched the transfer, not my divorce; that was still to come.)

We stand there dumbly for a while, then she claps her hands together. "Well! Now that you're finally here, there's a question that's been bothering us ever since we bought this place. You've *got* to know the answer."

She leads me from the verandah into the living room, exactly as I remember it including the big wooden daybed sofa and the glass-topped hollow tables with shells inside, arranged on a bed of white sand. In the Bahamian style, the exterior siding of the house actually forms the walls, and the studs are bare.

She points. "What in God's name are *those*? Can you tell me? We didn't want to take them off, of course."

Attached to some of the studs are small brass cylinders, with screw tops and two little tubes protruding from opposite sides. I take a deep breath and give Joan a lopsided grin.

"That's easy. Those were the air valves for the aquarium."

A Charming Couple

Mummy

1937

IN THE SPRING OF THE year after she'd met the man she wanted to marry, my mother set sail for England on the SS *Bremen*, chaperoned by her brother, Cummins, and his new wife. Following his magical summer in the States, my father-to-be had returned to London to work drearily as a stockbroker in his stepfather's office. She was rather shamelessly chasing him down.

My mother had an ardent Philadelphia suitor who'd proposed marriage. He was sweet, he was honorable. He was from a nice family. But there wasn't much of a spark. She told him she'd give him an answer that June.

Her first sight of my father in London was disappointing. "Charlie is less handsome than he was last summer," she wrote in the diary she'd started that year, "and has cut off his moustache, which is a mistake. City life does not become him. Am I a bit disappointed? Maybe a trifle. I don't quite know yet."

But the next day, "Charlie took me to dinner at a Spanish restaurant, and then to the movie 'Thunder in the City,' in which Edward Robinson starred. Later we went to a night club and walked home through the quiet city streets, which I enjoyed immensely. I'm no longer disappointed. He's grand!"

There are no more references to the Philadelphia suitor. She left London for a weeklong skiing trip to Switzerland but spent her time mooning. Back in London, now chaperoned by her aunt ("Aunt Arline agrees with me about Charlie's charm and urged me on to make haste!"), she accepted my father's invitation to visit his family home in North Wales. He kissed her on a picnic lunch in the moors above Llangollen, but "Charlie is unfathomable. I really can't make out if he loves me or not!"

Once again, things are different the next day: "Incredibly amazing. Charlie proposed to me this evening! I must be dreaming!! I did not give him a definite answer but I am thrilled and of course will say yes!"

My father's diary is more succinct. On proposal day he wrote: "Asked Louise a v. imp. q."

They were married at my uncle's house near Philadelphia on August 19, 1937.

Daddy

It was easy to see how much my mother loved him, and what she loved about him. I happened to be with her once when she was buying him a sweater, describing his build to the saleslady: "He has *very* broad shoulders." With a little smile and sparkling eyes, as if she were confiding a special secret.

My father's enthusiasms couldn't have been more appealing to her, and, later, to me: beautiful fast cars, handcrafted shotguns, speedboats, cliff-climbing, chopping down trees, bonfires, skiing, diving. He'd tell me fabulous stories about his childhood in North Wales—his own father away in the Great War (in which he was to die)—running wild through the countryside with his two brothers. The time he packed a ceramic ginger beer bottle with blasting powder from a nearby slate quarry and blew up a neighbor's predatory cats. Hanging by his hands from the top of a fifty-foot bell tower to impress his younger brother, Nigel. Shooting himself in the hand while demonstrating to his other brother, Patrick, that the air pistol was unloaded, then driving himself to the doctor at age thirteen because no adult was around. And, in his British Raj phase, shooting tigers in the Sundarbans, crashing his Gipsy Moth biplane, falling asleep late at night at the wheel of his Alvis, and waking up in someone's kitchen still in the Alvis. Adventure stories, frequently ending in fantastic, spine-tingling disaster.

How did he crash his plane?

"Some friends were in a speedboat, roaring along a canal. The drill was to swoop down on them and bomb them with a bag of flour. My boss was in the front seat, holding the bag. After he dropped it, I put the plane into a steep climb, looking over my shoulder to see if we'd hit them, and the plane stalled and spun into the ground. Neither of us was badly hurt, but my boss was pinned in the wreckage. Jolly lucky it didn't catch fire. Anyhow, I lost my flying license in India for good."

I'd sit there with my mouth hanging open in admiration. Were other kids' fathers this cool?

"I had my pick of anyone in Philadelphia," my mother used to say (meaning anyone in *polite* Philadelphia), "but they all bored me to tears. Then along came your father."

He was never academic. At Eton, England's hoariest and most tony school for boys, he was known as Stalky, Kipling's scheming miscreant from the generation before. One of his more flamboyant acts was to paint ammonium nitrate paste on the floor in front of an unpopular master's door late at night. When dry, this paste will explode at a touch, like nitroglycerin.

Sure enough, when the master stepped into the hall the next morning, there was a satisfying blast. The master survived, but my father was denied admission to the Royal Military College at Sandhurst and by extension to the posh family regiment (the 60th Rifles). Problem cases like my father's were customarily sent to the colonies to shape up, and so he duly found himself a shipping clerk in the British Raj, in India.

The clerkship had been arranged by Stanley Greenfield, a stockbroker in the City whom his father's widowed mother had recently married. Greenfield had a client who owned an import-export company based in Calcutta. The two knew each other socially and often went on shooting parties together. As a favor, the client agreed to take my father on.

Greenfield was a legendary wing shot, and in due course the client withdrew his account in pique after losing a shooting match with him. All favors to Greenfield were withdrawn, and suddenly my father found himself on sufferance in Calcutta with no prospect for advancement. To make matters worse, there was the plane crash in which his immediate superior had almost died.

The year was 1934. A wonderful adventure book had just been published, *Deep Water and Shoal*, about a round-the-world sailing voyage in the smallest vessel ever to make it up to then, thirty-two feet long. My father and an English friend in Calcutta with some resources read this book with great excitement.

"I have a plan which may surprise you," he wrote his mother in Wales. "Another man named Kettlewell, who also feels he cannot stand this place another minute, has decided to leave too and we are getting a 30 to 40 foot sailing boat and going to spend the next 3 or 4 years sailing around the world, don't laugh, its been done several times by people with as little experience as ourselves and in smaller boats. Anyhow, Mum, it seems to be the only thing worth doing that is left." He was twenty-seven at the time, and he begged his mother to continue his allowance. "I feel this is an opportunity that only happens once in a lifetime."

The two returned to England to drum up support and to find a boat. A year after their original decision, they embarked from London in an ancient yawl bought by the friend, the model of which now sits in my New York loft.

It took them two more years even to reach Barbados. There were storms. There were dalliances. There was a mutiny, in which my father was forced to throttle senseless a crewman they'd taken on in Gibraltar . . . a writer. Finally, they ran out of funds.

The boat was sold (for £375), his friend returned to England, my father bummed passages around the Caribbean and finally made his way to Bermuda in time for the finish of the Bermuda Race from Newport, American yachting's smartest event. Was this mere coincidence?

In the round of parties that followed, he met my uncle, Cummins Catherwood, heir to a coal-mining fortune, who'd raced his sixty-five-foot yawl *Valhalla* over with a crew of six. Some of his crew were flying home, and he needed able bodies for the return voyage. My father was only too happy to volunteer.

A big bash was held in Philadelphia for the returning sailors. My uncle Cummins told his older sister, Louise, thirty and unmarried, that she had to come . . . There was someone she should meet. An Englishman.

My father's family nickname was Bubbles, because as a child he had closely resembled the radiant boy in a popular soap ad. He was thirty himself at the time of my uncle's party, and how he looked then was a source of comment even by the throttled writer, one Eric Muspratt, who took his revenge in a "scandalous" account of the voyage titled *Going Native* (Michael Joseph Ltd., London, 1936). Muspratt cites the story of a doctor flirting with my father's widowed mother at a dinner party by flamboyantly describing the body of a young man he had examined that day: "the most perfect specimen who had ever come under my hands." She let him go on for a while before admitting this perfect specimen was her son.

His diary is uncharacteristically blank for the day after his meeting with my mother and for the next four days too, so there's no written record. I've tried many times to imagine it.

No great beauty, my mother, but she was always at her best at a party. Her face would be flushed, her small eyes almost hidden—"wreathed in smiles." She would ask all the right questions. She would know everybody and make all the right introductions. There would be a tasteful hint of coquettishness. But she would have already made her choice. My mother made decisions very quickly. And acted on them even faster, because, after all, how did you know if they were right? She was an existentialist while Sartre was still in short pants. She was also a romantic.

My father would look pleased and shy, but his eyes would be sparkling with tasteful mischief. His borrowed tux would not conceal the perfection of his body. He'd have a marvelous tan, his strong, brown long-fingered hands would move capably. He'd smell of Bay Rum. His body would progress through a series of graceful poses. His accent would be Eton. His talk would be anecdotal, full of amusing misadventure. His laugh—*duh-hunh*—would be accompanied by a cock of the head and would sound slightly conspiratorial. His teeth (most of the front ones false after the Alvis crash) would be wonderfully white and even under his clipped mustache.

The scandalous Muspratt had written: "Charles himself is very slow-going, very English. Nevertheless, I've kept on noticing that he is always very quickly on the spot when anything definite enough has to be decided or done. He gets there in one, though with a lazy, almost stupidly elegant and drawling fashion."

My father would have realized there was something definite enough to be done here. She was rich and he had no money, but it was more than that. He was a bit of a romantic existentialist himself: love at first sight would have been in character for both of them. To begin with, he accepted my mother's invitation to join her on a summer cruise aboard the *Valhalla*.

"Charlie Chaplin joined our boat in a dramatic way," wrote my aunt Ligi, who was also aboard, "arriving on a sea plane which landed nearby, taxied through the fleet of boats, and pulled alongside. He stepped out on a pontoon, threw his sea bag up on the high deck, grabbed a rope, and climbed hand over hand to swing on board. He was extremely good-looking, tall, and athletic . . . we all marvelled . . . and he distinguished himself on the roughest days of the race by climbing high up on the mast and changing the sails when the occasion demanded . . ."

I can see him now.

Hog Island

Buying In

January 1948

"We had the best of it," my father used to say of our thirty-year sojourn in the Bahamas, and from his point of view he was right.

During World War II, with the Duke and Duchess of Windsor presiding in Government House, Nassau entered a particularly lush period fueled by rich European refugees, Canadian tax exiles, and ambitious local white businessmen setting up the first fancy but tasteful real estate developments. Kneesocks and Bermuda shorts, plummy British accents, garden parties, gossip, polite scandal, and at least one spectacular unsolved murder (of the Canadian millionaire Sir Harry Oakes, complete with juicy rumors of the duke's involvement).

Acute gastritis combined with a change of citizenship had kept my father out of active combat, though he did return to England for a year to work as a welder in an aircraft factory. His youngest brother, Patrick, was killed in North Africa, serving with the family regiment, while Nigel, next up the ladder, was awarded the Military Cross for gallantry in the field as a cavalry officer in Burma.

This heroic legacy, along with my grandfather's death at the head of his battalion in the earlier war, must have weighed on my father a bit heavily. He never talked about it but he kept their portraits and medals in glass cases on top of the piano for all to see. By the end of the war, which he spent drumming up local support for England with the British Consulate in Philadelphia, his gastritis was bad enough to warrant admission to the Mayo Clinic. The doctors recommended he spend a month or two relaxing in a warm climate.

My uncle, Cummins Catherwood, had many friends in Nassau at the time, and a blind rental on Hog Island was easy enough to arrange. He told my father that the sea was clear as a dry martini.

Before the war, the 624-acre island, reachable only by boat from Nassau, had quietly developed into one of the most exclusive wintering spas in the hemisphere for American tycoons. Most of it was owned by Edmund C. Lynch, founder of the eponymous brokerage firm. But the most desirable beach, called Paradise, fronted on a narrow peninsula extending west opposite the Nassau town dock. A typical property was a cross section of this peninsula, featuring the curving, talcum-powder ocean beach on its north side and the harbor with excellent docking facilities on its south. Residents included Phillip Gossler, chairman of the board of Columbia Gas & Electric; Huntington Hartford, the A&P heir; and Arthur Vining Davis, president of Alcoa Aluminum. The Porcupine Club, two doors down from our rental (the Gossler house), had counted J. P. Morgan, Howard Hughes, Andrew Mellon, and Vincent Astor as members.

A Cheeveresque cocktail hour routine developed among these nine properties. The westernmost was owned by Hope Ryan, originally married to an heir to the Thomas Fortune Ryan transportation and tobacco fortune. After his death, she married a New York bon vivant named Cliff ("Baron Von") Klenk, charming and raffish. Like John Cheever's hero in his short story "The Swimmer," Klenk would slowly drink his way east along the peninsula, picking up followers along the way. The hardiest would arrive hours later at the Porcupine

Club, have one last planter's punch, and make their way home along the beach. Some swam.

My mother slid into Hog Island society as cozily as into a fitted glove, and the British colonials across the harbor in Nassau opened their arms to my Old Etonian father. The new couple in due course received the coveted invitation to the annual ball at Government House.

I was too young to take much notice of those early winters. We rented another house on Hog Island, farther to the west. Finally, at dinner one evening, my father tapped his knife on his wineglass with a twinkling, Christmassy look in his eyes: "I have an announcement. Hope Ryan has accepted our offer."

My mother looked to me as if she were tasting an unappetizing dish. The place Hope Ryan was selling, right next door (she'd inherited two properties from her mother, Mrs. Gardner of Boston), was an ancient wooden bungalow that needed a lot of work—nowhere near on a par with the house we were renting, which was not for sale.

The offer was for $25,000. My father argued that the low price justified extensive renovations.

My mother smiled with one corner of her mouth and took a drink of water. I didn't realize it at the time, but her money would be financing all this. "Well, it certainly needs them, doesn't it? Where are you going to begin?"

My father laughed. "I think with a decent dock, don't you? So we can land supplies."

What Does Home Mean?

January 1951

WHAT DOES "HOME" MEAN TO a six-year-old boy? It turned out not to be the big, gloomy house in Philadelphia where I'd lived until then. When my father and I arrived at the new place on Hog Island two weeks before the rest of the family to oversee the finishing touches, everything felt mind-blowingly familiar, as if I were coming back to somewhere I'd known forever. The year before, when I'd first seen the place, we were just looking it over. Now it was ours.

Touch and smell. My bare feet on the Abaco pine floors; the corrugated limestone rocks and fine sandy soil in the yard; the rough, warm cement of the walkway; my hands on the white-painted wooden settees along the sides of the deep north and south verandahs; the mildewy fragrance of the rooms; the salty easterly breeze in the corridors.

My room was in a small separate cabin, reached from the main house by a series of round cement stepping-stones in the light brown sand. I'd wake up before dawn, wander down to the dock in just my shorts, and watch the sun rise in the east over the flat-calm harbor. I'd be in a trance, not thinking anything at all, feeling the rough planks biting into my thighs and my pectoral muscles shivering in the cool air.

I wasn't looking forward to the arrival of "the Hens," as my father called them: my mother, my younger sister, her nanny Miss B, and our maid Grace. They would complicate a simple, delicious relationship: my father and me.

He seemed to feel that too. Didn't he? He was proud of my swimming skills (it turned out I loved waves, the bigger the better), the way my skin tanned quickly to a smooth molasses, the way I could shinny up palm trees to gather coconuts and walk carelessly along the top of the ten-foot-high pink stucco wall that bounded our property to the west. The way I'd learned to handle our heavy Abaco dinghy.

"Stop playing the fool, Gordy," he'd say fondly if my enthusiasm bubbled over into mere silliness.

But plumbers, electricians, and carpenters were finishing up the renovations, and inevitably the Hens showed up. "Fish-belly white," my father called them when they got off the plane, and I was proud to compare my tanned arm to anyone who would stand still for it.

When we got to the Chaplin House, I rushed around trying to show everybody everything. But I knew nothing would ever be the same.

My sister, Susie, and my mother had already started their lifelong battle. From my room in the little cabin behind the kitchen that I shared with Susie and Miss B, it was easy to hear them screaming at each other, a four-year-old toddler and a forty-year-old matron. A crash as my mother threw something breakable—a jar of cold cream?—against the wall. Miss B's low voice, as she tried to calm things down.

I can't remember ever using our joint bathroom. I would always pee outside, into the soft, brown soil around the shrubs. I must have pooed, showered, and bathed sometimes. But where?

Sometimes I'd overhear my mother giving instructions to Miss B, a trim little Englishwoman with twisted, arthritic hands:

"She is to have her enemas twice a day. Eight and five. Understood? You must make sure she finishes all her food, I know how she likes to hide it away. Don't let her get away with it! She should be brought in to us without fail for half an hour before dinner. We will come for her at other times when necessary. She is to be in bed, lights out, at seven thirty. If she wakes up before seven in the morning, she is not to be taken from her bed until then."

She was repeating patterns set by her own mother, an invalid who put her under the control of an evil governess as a young girl, to be marched in once a day for a parental inspection while they questioned the governess about her progress, never talking to her directly. But Miss B was kind and caring, while my mother's governess had been a sadistic German-Swiss who kept her on a choke leash, banning friends, censoring presents, denying sports.

I couldn't remember ever having had a governess, but then I couldn't remember my mother taking care of me either. She liked me better than she liked my sister, that's all I was sure of.

There was a wonderful piece of furniture in the living room of the Chaplin House (still there when I returned fifty-four years later)—half bed, half sofa, with an inclined back around three sides and lots of cushions. The design was Hawaiian: a *hikieae*, my parents called it. My mother and I would lie on it side by side after dinner, reading, while my father wrote letters at his desk in another corner of the room. Our legs and elbows might be touching. I'd think of the time, three years earlier when she was pregnant with my sister, we lounged side by side on her couch in Philadelphia, me rolling a toy metal truck down the hill of her belly and being unexpectedly offered a breast. For the rest of my life I'd associate that blissful moment with Susie's arrival.

A love of long perspectives was one of the things my mother passed on to me . . . dim hallways in hotels, piers, straight treelined roads, skyscrapers, bridges, plowed fields, flat country. Sometimes she took me on long, aimless drives or walks. We'd follow a road or trail, but it wasn't the place they led to that really interested her.

We almost died together on a trail. In late August on Mount Desert Island, Maine, a little forest fire began smoldering at Dolliver's Dump a few miles inland from Hulls Cove. Later it would grow into the Great Fire of Bar Harbor. My father was away on an ocean race to Mount Desert Rock. My mother and Mr. Collier, a raffish photographer who'd married into the McCormick fortune, decided to take their sons to have a look.

The trail climbed up on top of a ridge, and we could see the fire much closer than expected. In front of the flames was the familiar Maine landscape. In back of them were black cinders, smoke, and white, stripped tree skeletons, like hell itself. The line of demarcation was moving slowly toward us.

"Look, boys," my mother said, putting her hand on my shoulder. Her voice was alive and excited. Even her hand felt different.

Not too long after the first sweet-smelling tendrils of smoke curled past, we heard a crackling in the dry woods behind us, down the ridge where we'd come from, and saw that the topmost pine needles in the few dead trees were on fire.

"Oh, my lord," I heard Mr. Collier mutter.

He was a big man and easily picked up his son, who was about my size, and put him on his shoulders. His eyes jumped past us.

"Louise, we've got to run for it."

He left the road and charged downhill through the bracken and low shrubs toward a section of forest that the fire hadn't yet reached. In a few minutes, his son's blond head on top of his shoulders had disappeared in the trees.

"Gordy." Eyes straight into mine as if seeing me for the first time. "I can't carry you. You're going to have to run with me. Run as fast as you can. And don't be frightened. Just run and everything will be fine."

She couldn't have been smiling, could she? "Go *now*," she said, to keep me in front of her. And I went like the wind, feet barely touching the ground, as close as I could remember to flying. When I reached the tree line, I stopped and turned to see how she was doing. She was right behind me, flying too, like I'd never seen her. We both had wings. "Go on," she laughed. "Keep going. You're doing *gloriously*."

She loved telling the story to friends. "Sarge Collier just took off without a backwards glance and left us there," she'd say, her face red and lively. "But Gordy was wonderful. He never lost his head. He just flew. Didn't you, Gordy?" I'd blush in embarrassment and resentment: there it was again, that shimmering social voice, so different from her voice on the trail.

Glorious was one of her favorite words. So was *loathe*. . . . I thought she liked *loathe* better because of the way it rolled off her tongue, not so much because of what it meant, but I could have been wrong. Anyway, she used it a lot with my sister.

As in a steamy bathroom where nothing is clear, I could never remember exactly what my mother and Susie fought about. Whatever issue came up, basically. Maybe it all came down to this: I'd loved to nurse, and Susie had hated it. At four, she was blond, bright blue-eyed, and hellacious—a real screamer.

Now that the Hens were here, life at the Chaplin House began to settle into a routine. This included homeschooling by my father—in good weather on the north verandah overlooking the beach, in bad weather on the south verandah overlooking the garden.

My first-grade reader began "'Oh my,' said Jane. 'See Spot run.'" My father reacted right away, taking out a red pencil, crossing out the "Oh my," writing in "Odds bodkins." Same thing all the way down the page: "gadzooks," "by my halidom," "Gorblimey." A revelation! A world of new possibilities opened up, and I think it was at that moment I became a writer.

Finally, he threw away the first-grade reader and began to read me chapters from his own childhood favorites: *The Adventures of Robin Hood, William the Outlaw*, *Treasure Island*, et cetera. But only a chapter of each. If I wanted more I had to learn how to read it myself, so I did. One of the books was so scary I took it out and buried it in the sand.

When he presented me with Marjorie Kinnan Rawlings's *The Yearling*, with illustrations by N. C. Wyeth, I felt an intense shock of recognition. The Rawlings house was on a piece of high ground in the Florida backcountry, an "island," though there was no water around it, just an ocean of scrub. But in the Wyeth illustrations the Rawlings house looked exactly like ours: a simple white clapboard bungalow with deep verandahs.

My mother would be on the phone, setting up luncheons, picnics, teas, dinners. Or just gossiping. The way her voice would change into its lively social mode marked the start of my lifelong neurosis about exclusion. Her friends took precedence. At parties, in the English tradition, my sister and I were expected to be seen and not heard. And not seen very often.

Sometimes in social situations she used that vivacious tone with me, but it always rang false. "Gordy, could you be a dear and run back to the house and fetch my sunglasses?" Sitting on the Porcupine Club beach before lunch with friends. Smiling at me, but talking for them.

A quarter of a mile to the house, find the glasses, and a quarter of a mile back. In a blind rage, I ran the whole thing as fast as I could, pretending to be dying of heat and exhaustion by the time I got back to the group of women, and hurling the glasses in my mother's amazed face just before I collapsed.

The Aquarium

February 1951

IT WAS EXACTLY LEVEL WITH my eyes, on a table in the big living room with a fishy, underwater backdrop painted by my father. Water straight from the ocean, a bottom of white sand, a few rocks, an empty conch shell, and a little pump with a soothing hum that blew slowly rising silver bubbles.

My father had discovered that the high, rocky spit protecting the east end of our beach contained several large tide pools. At first glance, they seemed empty except for a few common black-and-gold-striped sergeant major damselfishes in mid-water and pale, almost transparent Molly Miller blennies sitting on the weedy bottom.

"Why are they called Molly Millers?" I asked him.

The little fishes had long, severe profiles and down-turned mouths. "Because they look haughty and disapproving."

"Oh."

We were both wearing sneakers to protect our feet from the sharp, spiky limestone that had been coral a few thousand years earlier.

The idea that all of the island, all of Nassau, all of the Bahamas had once been coral, was unbelievable. From the top of the spit, I could see dark reefs of it off to the northeast. Would those form an island someday, just growing out of the water? What if coral took over the ocean and turned it all to dry land?

My father moved deeper into the tide pool, where I could see a large, flat rock on the bottom.

"All right, Gordy. Now watch carefully." He curled his fingers under a corner and strained, muscles standing out in his back and along his ribs.

The rock tipped, dripping, out of the water, revealing a population of living things that seemed straight from a dream. The starfish had long arms like tentacles, some dark and covered with short spines, some smooth and ringed with green-and-white stripes. They moved these arms rapidly to push themselves into the water, along the bottom, into other crevices.

A couple of fast-swimming blue-clawed crabs with fins shot away like fish. A small octopus, body the size of a Bahamian lime, changed from red to white to dark brown as it sashayed over the bottom to slip into a crevice. Two red-and-white-banded shrimp with glowing green eyes and long white feelers stepped carefully around a cluster of tiny black-and-white-striped sea urchins, rapidly wiggling their spines. We caught a glimpse of an incredibly beautiful speckled eel, about which my father would write fifteen years later in *Fishes of the Bahamas*:

"The light portions are pale to golden yellow, the golden color especially pronounced at the tail tip, where it is most extensive. The dark portion is brown to purplish black. This is probably the most beautiful of Bahamian morays but, while common, it seldom is seen on the reefs because of its secretive habits."

Sure enough, the eel disappeared in a split second, so fast I could hardly believe it had been there at all.

My father gently lowered the rock and we stood there staring at each other. He had a big smile, almost like a kid himself. I had never felt closer to him. We were exactly on the same wavelength.

"Wow," I said. "How did you . . . ?"

A few days later he came back from town with the aquarium. After we set it up, we went back to the tide pool with a bucket, dip nets, and a small seine, which we arranged around the rock. Catching the creatures under it was wonderfully exciting, but the beautiful eel (a goldentail moray) wasn't in residence that day. We did get the octopus and the two shrimp.

Counterintuitively, my father named the shrimps Hermann and Adolph and the little octopus Winston. Winston would lie in the mouth of the conch shell all day, plotting and planning, his breathing siphon expanding and contracting, his green eyes cold, depthless, intelligent. One morning there was one less candy-striped shrimp.

"You don't want to get on the wrong side of Winston, small as he is," my father said. "Look at what happened to poor old Adolph."

To save the remaining shrimp, we took to feeding Winston little pale beach crabs that we'd catch when they came out at night to run over the sand. Winston never failed to electrify. At first he'd seem to ignore the crab as it wandered around, sometimes even stepping over him. His color scheme would blend perfectly with the conch or wherever else he happened to be lying. Except for his pulsing siphon he could be a rock himself.

Then you'd see him gather himself together, his tentacles slowly coiling beneath him, like a cat getting ready to spring. He would turn darker and seem to grow in size.

The doomed crab would continue to act oblivious, wandering across the sand, maybe even stopping to pick up a morsel with its claws. Its eyes would stand up straight on their stalks, its mouth parts would move gently, almost as in speech. Time would stand still . . . like that moment just before you jump off the high dive or take the drop on a roller coaster. Watch out! WATCH OU—

So fast! It always made me jump, goose bumps popping out on my arms and legs. Suddenly the little crab would disappear. In its place would be Winston, blushing blood-red, eyes still cold and glacial. Sometimes a crab leg would protrude from the web of tentacles, but mostly nothing would show that the crab ever existed.

"W-w-what happens now?" I stuttered after the first session.

My father spoke with relish and fascination: "First Winston injects it with a digestive poison that kills it and softens it up. Rather like a spider. Then he'll eat it."

"What does he eat it with?"

"He has a beak."

"Where?"

"In the middle of his tentacles."

"Wow! What does the beak look like?"

"A bit like a parrot's."

"A *parrot's*?" One of our neighbors on Hog Island had a collection of scarlet macaws. Their huge black beaks could strip a green almond nut in seconds.

My father smiled. "But smaller, of course. Very difficult even to see it."

"How big do octopuses grow?"

His smile broadened. "I'll give you a book to read about that."

The book was Jules Verne's *Twenty Thousand Leagues Under the Sea*, featuring an octopus the size of a city bus, but he assured me it was just fantasy. Real octopi didn't grow to be more than six feet from tentacle tip to tentacle tip. And they were only to be found in Puget Sound (wherever that was).

He told me that octopi were so smart and sensitive they could have nervous breakdowns just like human beings.

I couldn't help noticing there was never anything left after Winston had finished his meal. Not a single sliver of crab remained.

Life in the aquarium was mostly peaceful. Each fish, shrimp, sea urchin, or starfish had its own comforting routine, and every one of them looked contented. Even happy. The sergeant majors swam about unconcernedly and a bit stupidly, while the blue-and-yellow beau gregories were territorial, always chasing each other away. Nothing serious, though. It was more like a game.

My favorite fish was the male bluehead wrasse, a present from Mr. McKinney, who'd caught it in a fish trap out on the reef. We'd never seen one in a tide pool. This fish was the ultimate in graceful elegance. Its looks were striking: a dark blue head and blue eyes, a black-and-white-banded ceremonial belt at the waist, and a deep green tail feathering out to two long points, with a translucent webbed membrane stretched between them. Like a mermaid's.

But the way it swam was what really got me. Instead of using the tail like other fish, it used its two pectoral fins like a swimmer's arms, carrying its tail languidly behind. If it twisted or changed direction, the tail would follow, like a cat's. Sometimes it would sink to the bottom, resting lightly on the tail, but mostly it described sensually sinuous arcs, its small, fine mouth opening and closing as if in song. Only in an emergency—if being chased by a beau gregory, or if a piece of fresh conch were dropped into the aquarium by my father—would it bother to use the tail, and then it could move so fast you couldn't follow it. That was the way I wanted to move myself.

Two a.m. I wake up weirdly alert. A norther is coming through, and the palms outside my window are clattering in a cold, raw wind. My first storm in this new place. Without really thinking, I put on shorts and a sweater and go outside. There's no one else around. They must all be asleep.

It's almost a full moon, and clouds are rushing south, making it seem as if the moon is sailing forward into the wind. The white buildings darken and lighten in the cloud shadows. Stepping carefully over the cement walkway (my father had caught a large tarantula here a few days ago, and had told me a spine-chilling story about a fugitive hiding out in a dark room and hearing tarantulas clicking their fangs), I head for the living room. I want to make sure the aquarium creatures are okay.

The whole house trembles slightly in the wind. Out the picture window facing north, the moon illuminates a churning whiteness. Feeling my way across the dark room, I flip on the little aquarium light. Amazing! The starfish are writhing rapidly over the sand or climbing the glass panels. The banded shrimp, Hermann, eyes aglow, is positively on promenade. But the bluehead wrasse has disappeared and Winston is strolling lightly on the tips of his tentacles, en route to where? Or is he just dancing? Anyway, he looks ecstatic. Suddenly, he presses his tentacles together straight behind him and shoots through the water toward me like a fish, landing on the glass panel in front of my face. I pull back quickly. I can see each sucker cup as it clamps on to the glass, but his beak is still hidden in the soft, white area where all the tentacles come together.

He's eaten the bluehead wrasse—that is obvious. Winston is a dire threat to this peaceful world.

He has already climbed out of the tank once in the night, to be found half dead under a chair. Suppose he climbed out again . . . for good? Or so it would seem. If I could get him to go into his conch shell, I could carry him down to the harbor and dump him in, replacing the shell afterward. No one would be the wiser. He'd be happier back in the sea.

I take the dip net from its shelf under the aquarium and pull up a chair to stand on. When I gently poke Winston with it, he turns dark brown and curls his tentacles around his body but stays where he is.

I poke harder. Suddenly a tentacle whips around the dip net and I can't pull it free. I yank harder . . . a second tentacle lassoes it. His cold eyes shift. Can he see me through the glass?

The net is now useless. I can barely wiggle it. By extension, I feel a bit like a captured crab. Winston has grown in size, like he does just before the coup de grâce.

Something snaps inside my head. I drop the net, plunge my hand blindly into the tank, close it over Winston's soft body, and squeeze as hard as I can, gritting my teeth as his tentacles pluck at my skin. I'm waiting for the sting of the beak, but it never comes. By the time the tentacles finally go limp, I'm in tears.

But I've *won*. My wrasse is avenged and the underwater world made safe. I dispose of the corpse out in the garden and lie in bed shaking for what seems like hours, listening to the wind.

Next morning at the picture window my father and I watch huge breakers pound the outer reef, tops blowing back like horses' manes. Seaweed and other detritus bowl along the beach in the gale. Casuarinas thrash like underwater plants in a strong current. I don't dare to look at the tank.

I stand there glued to the window, acutely aware of my father pottering around and finally approaching the tank. My ears burn. After a minute I hear him say: "Oh, dear."

"What?" I turn from the window to look at him as casually as possible.

"Winston's on walkabout again." He starts searching the floor, peering under pieces of furniture. "The storm must have stirred him up, poor thing."

"Maybe we should put something over the top of the tank," I say.

"Excellent idea, Gordy. We will do that forthwith. Help me look for Winston, will you?"

First, I go to the tank. The bluehead wrasse is there, describing its usual sinuous arcs. Back from the dead. Goose bumps pop out on my arms. I can't believe my eyes.

Careful study, later on that day, reveals that blueheads, like other members of the wrasse family, bury themselves in the sand at night. Some wrasse, the razor fish, can actually swim under the sand. I killed poor Winston mistakenly, and I feel pretty bad. But at least my intentions were honorable: I'd do the same thing again in the same circumstances.

Totems

February 1951

A CALM, WARM MORNING IN MIDWINTER. My father decided it was time to fit me with a face mask and introduce me to the "silent world," as his hero Jacques Cousteau would christen it.

I waded out from the beach until the silky cool water reached my chest, and then I bent my knees until my head was below the surface. As if I'd passed through the looking glass like Alice, I was suddenly inside our living room aquarium with its colony of bright, tiny sea creatures.

My smiling, begoggled father was beckoning in dreamy slow motion. Easing through the silvery ceiling of the sea, through clouds of minnows, over the dancing white sand bottom, I swam out with him until the world changed from bright sand to beige-colored rocks fringed with plants and set with purple and yellow sea fans.

My father dove eight feet to the bottom, where I could see a little cave underneath a ledge, and beckoned again. Diving down to him was as easy as flying. Under the roof of the cave, a living jewel was hanging upside down, shading from deep purple at its head to brilliant yellow at its tail. It turned sideways with a wave of a magenta fin and cocked a midnight blue eye.

Click. Something like a mild electric shock went through my head, and the entire scene—the bright aquamarine water, the dark cave, the overhanging brown rock, and the luminous little fish—took on the kind of warm glow I'd come across much later under the influence of hallucinogens. An old friend described a similar childhood experience involving a yellow warbler sitting on a purple thistle growing from a green lawn close to a white clapboard house. On the strength of that image, he was to become a fanatic birder.

My totemic fish was called a fairy basslet, my father told me when we came up for air. It was the catalyst: without it, the scene would not have come together. Was there any way we could get one for the aquarium? If we could, I felt sure that everything in it would become magical. And I'd be part of it.

A few weeks later, on a picnic to Treasure Island, I encounter the flip side of this coin. My mother and the other women are skinny-dipping near a talcum-powder beach in a little cove, while my father and the other men have disappeared on some fishing adventure. Wearing my new face mask, I float in the warm, clear water near the women, watching my mother's hands move in a caressing hula and noticing that her breasts are by far the most shapely of them all (and there are a lot to choose from). The women are engrossed in gossip, oblivious to all else, including the attentions of a five-year-old.

Finally bored, I drift away and begin exploring a shallow stand of elkhorn coral on the other side of the cove. My first solo flight . . . I have the silent world all to myself, though it isn't silent at all. Snapping shrimps pop and crackle, parrot fish crunch on the coral, and a light hissing like a kettle coming to a boil overlays everything.

Spreading branches of elkhorn reach eight feet from the sandy bottom to the surface. Their color is a light, glowing terra-cotta, the texture deeply serrated with the chambers of the polyps that had made them. A little school of bluestriped grunts hangs in the branches.

Through the clear water's prism, everything is magnified and intensified. The reef fish seem to be lit from inside—stylish, dark gray French angelfish with their down-turned white mouths, yellow-ringed eyes, and gold-tipped body scales; gaudily impudent, young turquoise-spotted yellowtail damselfish; lazily graceful slippery dick wrasse; pony-like, dainty tangs; blue clouds of chromis. The fish, anemones, purple gorgonians, soft corals, tube sponges, and sea fans all move to a light, watery rhythm, the symphony of the reef. I feel part of it all in a way I never did on land.

"Why did man ever come out of the sea?" my father loved to say, and I could see his point.

But being part of it also means being part of the indigenous food chain. I forget this until I feel a prickling among the short hairs on the back of my neck and turn to look behind me. An enormous, dark barracuda is hanging motionless two feet above the sandy bottom seaward of the little reef. It's shaped like a submarine and is significantly larger than I am. The undershot jaw is the most nakedly aggressive thing I've seen so far in my short life. The eye is huge and empty.

Without moving a fin, the predator turns slowly toward me and angles up until it's a circle bisected by lips and teeth. I panic and begin to thrash away toward shore, expecting any second to feel those jaws surgically remove one of my legs. The fifty-foot swim seems to take a lifetime.

Safe on the hot white sand, I throw off my flippers and face mask and wait in weird mute fascination for the fish to strike the bathing women on the far side of the cove. When nothing happens, I run screaming toward them: "Barracuda! BARRACUDA!"

Soon enough, I'm surrounded by nude women with wide eyes. "How big was it, Gordy?" asks the attractive Mrs. Sparrow.

I throw my arms wide apart. "Bigger than me!"

"My God." My mother's face is flushed and excited.

"What was it doing?" Mrs. Wanklyn asks.

I shake my head. "It was—"

"Did it come after you, Gordy?" asks Mrs. McKinney.

"Yeah! It did! It was going to come after me."

"Good Lord," says Mrs. Weingart. "You must have been scared out of your wits." She puts her hand on my mother's arm. "Yet his first thought was to warn us."

"Heroic!" says Mrs. McCutcheon. "Wait till Charlie hears."

In My Element

January 1952

AT SEVEN YEARS OLD, I may have been the youngest person ever to use SCUBA gear. As soon as it was available to the general public after its invention in France by Jacques Cousteau and Émile Gagnan, my father had snapped up several sets of the exotic-looking equipment: regulators and mouthpieces with accordion rubber tubing, steel air tanks of different sizes, harnesses. And after a lesson or two, here I was soloing slowly over the turtle-grass flats near our dock, one step closer to being a fish myself.

The regulator made a comforting hiss as I breathed in the air, and bubbles of exhale blew merrily out of it behind my head, shimmying to the surface like silver jellyfish. My breathing fell into a comfortable rhythm . . . in, hold, out, hold. I felt in my element.

With slight negative buoyancy, I settled on my belly on the sand in front of a ledge at the start of a plateau of turtle grass and marl. Black-and-white-striped slippery dick wrasse zoomed past, inches from my face; a tiny sharp-nosed puffer with a mouth like a miniature horse's hung motionless in the last slant of light before the dark ledge, feathering its pectoral fins.

There was something moving in the darkness way back under the ledge, and I slid forward carefully to avoid stirring up the fine sand. Slowly, I picked out a tangle of black-and-yellow-striped legs, thorny antennae, and glowing green eyes on stalks protected by horns. A nest of baby lobsters, you couldn't begin to count how many.

The tide was coming in. As I ventured out over the marly bottom, I could feel the current gently urging me east, farther into the harbor. The surface looked farther away than before.

Ahead of me, I could see a large rusted-out iron pipe . . . perfect for eels, lobsters, cardinalfish, squirrelfish, groupers, rockfish, banded shrimp, snapping shrimp, mantis shrimp. The first rust hole indeed turned out to be the home of a big green moray, with a body as thick as my arm and teeth like white needles as its mouth opened and closed with its breathing.

The pipe angled away from shore, and by the time I'd inspected every hole and crevice I was well out into the current and about twenty feet down. My legs streamed out behind me as I gripped rocks and ledges to keep my place, and the purple gorgonians growing from the bottom around me waved as if in a breeze.

A little way up-current I could see a large, stately indigo fish coasting toward me, hardly moving a fin. Its orange eye swiveled round to take me in, its wide streamered tail moved slightly to propel it to the side, and it drifted past so close I could almost touch its toothy green beak. The heavy, armor-like scales on its sides had light blue centers, and there was a light blue bridle below the beak. The beauty of the fish was hypnotic. Perfectly natural to let go of my rock and drift after it.

The current picked up in the deepening channel, and soon the big indigo fish and I were flying over the bottom like the spacemen in my father's copy of H. G. Wells's *The First Men in the Moon.* A leap combined with a somersault carried me twenty feet, but the startled fish shot sideways with a flick of its tail and went quickly out of sight.

The only fish I could see now were a little school of yellow-and-blue-striped French grunts, fighting the current to stay close to a head of brain coral. A glass-bottomed water taxi puttered overhead thirty feet up with a circle of surprised tourists' faces staring down. I gave them a jaunty wave.

I was fine, but maybe I should head back. Turning into the current and swimming as hard as I could, I could barely stay even with the brain coral head. What now? I was a little out of breath, but not scared. This was the harbor, after all: there were no unfriendlies in here.

The brain coral was slimy but sharp when I gripped it. Maybe the tide would change before I ran out of air. And maybe not. If I surfaced, someone would pick me up. But I didn't want to surface until I had to. I was having an adventure. I let go of the coral, and the current swept me gloriously away.

Somersaulting, leaping, flying. Free as a fish. And as thoughtless as one. No harm would come to me: strangely enough, the idea never even crossed my mind. I had a strong feeling I was lucky, and my father was the talisman. If it came to rescue, I knew he'd be there, but rescue was the extreme. It was enough to know he was around. A world with him in it would never hurt me.

After a while, I picked up the bubbling rattle of an idling outboard. A face looked down at me through a glass-bottomed bucket—my father's face. Joyfully, I started up toward it, eager to tell him about the green moray, the spectacular indigo fish (whatever it was), the space leaps . . . the whole saga.

He grabbed my tank from behind while I shrugged out of the harness, slid my mask up on top of my head, and looked up with an eager smile. His own mouth was tightened into a little purse of annoyance under his mustache. He hoisted the tank over the side and silently put down the swim ladder while I removed my fins to clamber aboard.

"You silly little fool."

Now I watched his hard hands, waiting for a crack on the ear. (The year before, when I got one, I pretended he'd ruptured my eardrum.) "What the bloody hell did you think you were doing?"

There was something unconvincing about his anger. I had seen him really angry only once, when Cliff Klenk, the new husband of our neighbor Mrs. Ryan, had clapped him on the shoulder during the usual cocktail hour hegira and exclaimed: "Gold diggers together, hey Charlie?" Klenk was never welcome in our house again. But this was a strange, almost reluctant kind of anger, so I knew my mother was behind it.

My mother never disciplined me herself if I misbehaved—shooting our maid, Grace, in the bottom with a BB pistol, teasing my sister until she cried—but would just say, "Wait till your father gets home." He'd come for me with those hard hands, pursed mouth, reluctant eyes: *CRACK*. Physical punishment came naturally to him. He'd been caned regularly at Eton.

The dock was about half a mile away—I'd drifted a good distance—but I could see my mother standing on it, shading herself with her red-and-white-striped parasol. My father put the outboard in gear, and I sat on the middle seat silently facing him.

I handled the bowline as usual when we arrived at the dock, and he hoisted out my air tank and weight belt. My mother just stood there in the shade of the parasol, eyes hidden behind white-rimmed, up-slanted sunglasses, like Auntie Mame's. She was wearing a green-and-white paisley print bathing suit with short legs and a pleated bosom.

"And how is your son?" she asked coldly.

"All right, darling," my father said.

"Does he know the trouble he's caused? Does he realize how worried we were? Does he even *care*?"

"Yes. I believe he does."

"He could have *died*. It would have served him right." I watched her frozen face in confusion, thinking of that time on the trail near

Bar Harbor when we *did* almost die together. Then she'd seemed galvanized, vital.

"He was all right, dear." My father sounded a little tired. "He was doing quite well, in fact."

"You must have been crazy to let him dive alone. Absolutely insane. The child is only seven years old."

My father's eyes shifted to me. There was something there, a tiny hint of . . . what? Complicity?

My mother must have seen it, too. She turned, walked quickly away up the dock, and disappeared under the palms at the beginning of the walkway.

"For God's sake, never do that again," my father said. "You see how upset she gets, don't you?"

"I didn't mean to upset her. It just happened."

"It just *happened*? What on earth do you mean?"

"This fish came by and I followed it."

My father stared at me and gave the laugh—*duh-hunh*—he used as comment on dizzying stupidity. "You'll have to do better than that, my lad."

"But . . . Daddy. That's what happened. It looked so nice."

"How did you think you were going to get back in that current?"

"I . . . didn't think about that part."

"*Bloody* hell. You must learn to think about that part if you ever want to go diving again. Don't just assume that someone's going to rescue you, for heaven's sake. What if I couldn't have found you? What if you surfaced in front of a boat and they didn't see you?"

Okay, I'd pay attention, if that would get me back underwater. Other than that, I didn't see what the fuss was about.

My father began to hose down the dive equipment. "What kind of a fish was it that you *had* to follow?"

"A parrotfish, but I don't know what kind. I've never seen one like it before."

"How big was it?"

I held my hands up three feet apart. "About like that. It looked *heavy*."

"What color?"

"Dark blue. Kind of violet."

"A midnight parrotfish." He rubbed his chin. "That's interesting. I've never seen one in the harbor."

"Where do you see them?"

"Usually on deeper reefs where there's plenty of coral."

"This one was just kind of drifting with the tide."

"Really? And what else did you see?"

"There was an old rusted-out sewer pipe or something with lots of stuff. A *huge* green moray." I swung my arms wide. "A little high-hat, too. Maybe we could scare it out into a net. It'd be good in the aquarium, wouldn't it?"

My father looked thoughtful. "If the beau gregories didn't chivy it about and nibble off its fins. The little brutes."

Eel Bay

February 1952

WE'D BEEN COLLECTING FISH IN tide pools, fish traps, and seine nets, but there always seemed to be many more interesting ones that were out of our reach. I was very excited when my father announced he had obtained an organic fish poison called rotenone, made from the root of the jicama vine and its relatives, and used by the Indians of the Amazon basin to catch food fish.

But it was a shock to learn that the fish we caught this way would be dead. Couldn't we use a little less poison so they'd be just knocked out for a while?

No, they'd be dead for good and all.

How many fish would it kill? My father wasn't sure. Only the ones in the immediate vicinity, he thought.

And how large was the immediate vicinity? He really had no idea.

It bothered me a lot. If you couldn't enjoy the fish and see how they lived, if all you had was a dead body, then what was the point?

My father explained that you would have a specimen, carefully preserved, of a fish that you'd never see otherwise, possibly that no one had *ever* seen, and that it could be studied by generations of scientists as part of a historic collection. I had to accept that, but I always imagined how the specimens would look if they were alive and swimming in our tanks. Especially the fairy basslets, which we never had been able to take alive.

We started in the shallow water off beaches and turtle-grass flats and—my father was right—collected many creatures we'd never seen before. The eels were particularly spectacular: black-and-white reticulated morays with rounded teeth; blind, pink worm eels, which spent their lives under the sand; snake eels with gold spots surrounded by dark brown rings on a pale green background; snake eels with two black stripes alternating with white down the length of their bodies; snake eels with one black stripe, a pale belly, and an orange top. The snake eels all had sharp-pointed tails and could insert themselves tail-first into the sand and suddenly disappear.

Rose Island, a couple of hours east of Nassau by boat, is a Felliniesque venue for collecting, so idyllic it borders on the surreal. A sheltered cove on the north shore with a half-moon beach backed by thick casuarinas, no other humans or human habitation in sight. The collecting team numbers seven: my father, mother, and sister Susie; my mother's old friend, Dr. H. Radclyffe Roberts, director of the Academy of Natural Sciences in Philadelphia; his no-nonsense English wife, Hazel; and the acting couple, Hume Cronyn (an avid fisherman) and Jessica Tandy, who looks sweet and beautiful even to nine-year-old me. We have come out in the *Dollie*, our bumboat of choice, run by a salty Spanish Wells captain named Preston Sands, who loves my father.

We wade in to the beach with our picnic lunch in a Fortnum & Mason wicker basket: three thermoses of grouper chowder, lettuce and mayonnaise sandwiches, cheese, fruit, white wine for the grown-ups, and sodas for the kids. The men lay it all out on a red-and-white-checkered cloth in the shade of the casuarinas, and the grown-ups sit down while my sister and I dawdle in the shallows watching thick schools of silvery pilchard circling over the turtle grass.

I can hear Jessica Tandy exclaim, "Ah, heaven!" as she tastes a spoonful of chowder from her bowl. My mother smiles happily. She is in her element.

After lunch, the men spread pillowcases of emulsified rotenone on the shallow turtle-grass flat near the beach, and soon enough strange creatures begin to surface. The eels are numerous, including yellow-and-white burrowing spaghetti eels with their protruding lower jaws, along with yellow-and-blue-striped grunts, pilchard, and the smallest wrasse in the Bahamas, the tiny green, red-dotted *Doratonotus megalepis*, so well camouflaged you never notice it without poison.

The cloudy, rotenone-filled water reaches just below my knees. I find the best way to collect fish in it is to shuffle along, feeling them out with my toes, so I ignore my mother's order to wear sneakers. (Susie's too young to follow suit.) When something like a very powerful, sharp-toothed mousetrap takes sudden hold of the second toe on my left foot, I scream and fall over backwards in the water, raising my foot clear.

"Gordon Chaplin was bitten quite severely on the toe by a green moray," my father's field notes read inaccurately (one of the few times I catch him out). To this day, the image is indelible: a fierce brown-and-white-speckled moray that squirms and flails from my toe, its needle teeth sunk deep into my flesh like hypodermics. The text in *Fishes of the Bahamas* on this family notes that "nearly all have strong jaws and fang-like teeth . . . [the larger ones] are potentially dangerous to divers."

This one is about two feet long, its muscular body covered with a coat of mucus that I can't get a grip on no matter how hard I tug and squeeze. Do I see Hume Cronyn stifle a grin? I must look a little ridiculous lying in the shallows yanking on this eel, but I'm scared out of my wits. *I can't get it off.*

"Let it go! Stop pulling!" a man's voice shouts.

Someone is cradling my head out of the water while my father works on the eel. "The brute . . . ah, the brute . . . A knife . . ." When Preston wades over with the ship's utility Ka-Bar, I imagine my father severing my toe to free the rest of me, and close my eyes.

After a minute or two I feel myself lifted into the air and carried through the shallows while women's voices mingle with men's in expressions of shock and concern: *My God. Very nasty.* My toe feels numb, but the weight attached to it is gone. Then I'm hoisted onto the *Dollie*'s transom and lie there like a boated game fish. I open my eyes when my father puts a hand on my forehead: "Poor old thing. How do you feel?"

"Okay. How did you get it off?"

"Pried open its jaws. Jolly hard, I can tell you. A real death grip. We're lucky the brute was small as it was."

"But the moray bite be poison, you know," Preston says. "I seen a mahn lose they hahn."

"Not actually venomous," my father says quickly. "But certain morays might have a toxic mucus in their mouths. It's never been scientifically pinned down. . . ."

"We need to disinfect that toe thoroughly," Mrs. Roberts says in her no-nonsense English voice. "Preston, do you have soap, fresh water, and a washcloth?"

"Yes, mam." Preston ducks into the *Dollie*'s cabin. I notice the wooden planks under my wounded foot are red with blood but can't bring myself to look at the toe. When he comes back with the goods plus a large aluminum pot, Mrs. Roberts lathers the cloth and firmly takes hold of my foot. "This will hurt a bit, Gordy. But you're a big boy now, aren't you? I've treated a lot worse than this in London during the war."

I close my eyes again and clamp my teeth together as the washcloth rasps into and around the lacerations. "He'll need some stitches, of course," Mrs. Roberts says busily, "to get that flap back in place. I doubt whether the nail can be saved, but the toe itself should be all right as long as there's no infection."

Finally, she drops the cloth into the pot and smiles down at me. "There we are, then. That wasn't so bad, was it?"

I shake my head and try to grin.

"And we have plenty of alcohol for the fish, luckily. Preston, can you clean out this pot? Now, Gordy, the alcohol will sting a bit, too," she says to me. "But you've survived the worst with flying colors."

"An honorable wound," my father says. "Received in the line of duty."

"In the name of science," Dr. Roberts says.

"You'll carry the scars proudly, my boy," Hume Cronyn intones.

Preston hands the clean pot to my father, who fills it half full from the polyethylene bottle of denatured alcohol used to fix the specimens, and sets it on the transom beside my foot. He takes hold of my ankle. "In with it, then."

The alcohol feels like boiling oil. I howl, tears shooting from my eyes while my father holds my foot down and Mrs. Roberts counts to five slowly. Then he lets go.

My mangled toe is red against the white paint of the transom. I can see holes in the nail where the teeth have gone through, and it seems to be slightly off-center. Blood seeps out from under the nail and from the torn flesh farther down the toe.

"*I'll never be the same again*," I think with some pride.

When I look around, I see my mother is watching me with an inscrutable expression from across the *Dollie*'s cockpit, and I realize I've been waiting for her to say, "I told you so."

In my honor, my father named the place Eel Bay.

The Cutting Edge

March 1952

BESIDES A GOOD EYE, MY father had an instinct for creative innovation. He'd acquired the Aqua-Lung (as it was called in the early days) as soon as it was available. He was using rotenone long before it became a recognized scientific tool. His career's eureka moment was to put them together.

Around noon on March 25, 1952, my father, Dr. Roberts, and I (my toe still heavily bandaged, but pronounced okay for immersion in salt water) took our outboard runabout west in the protected shallows behind the offshore islands and reefs to an isolated coral head near our favorite tide-pool collecting location, Delaporte Point. We were equipped with two aqualungs and plenty of rotenone. Conditions were perfect: low tide, a cloudless sky, a light south wind, and the usual martini-clear water.

The little coral head formed an underwater cove, completely protected from the current. My father and Dr. Roberts soused their pillowcases of powdered rotenone in seawater until it was fully emulsified, then donned their aqualungs and took it to the sandy bottom twelve feet down. I watched from the surface with mask and snorkel.

I'll have to admit the idyllic scene—schools of blue-and-yellow-striped grunts and dark blue, spade-shaped tangs cruising leisurely along the reef's edge while smaller reef-dwelling chromis, wrasse, and parrotfish arced through the coral valleys—only registered in passing. I was too excited to feel concerned about the fishes' fate. I'd picked up the charge from Dr. Roberts and my father: this was a first. We were breaking new ground. Maybe the rotenone would dissipate harmlessly out here, but nobody knew for sure.

Carrying a set of long-bellied collecting nets, one for each of us, I watched the two men slowly fin their way into the little cove and split up, one on each side. Rotenone was easy enough to disperse on the surface—you just poured it out—but what to do underwater? Dr. Roberts was trying to push the light brown clouds from his pillowcase into the reef by flapping his hand. My father hit on the idea of flapping the stuff in with his swim fins. That worked better. The crystalline cove slowly disappeared into a brown fog.

About ten minutes later, while they were still applying the last rotenone along the edges, "an absolute slick of stunned fish appeared," as my father was to write in his field notes. "Mostly grunts, which floated away on the surface. A fair-sized Nassau grouper (*Epinephelus striatus*) was knocked out, and several tangs. This was the first time we have tried rotenone in open water. It works very well. One needs a constant supply of nets and a dinghy standing by with buckets for large and small specimens."

Standing by with the dinghy was my job: I dove down and took the collecting nets as the two men filled them up, emptying their specimens into buckets along with my own collections from the surface (a whole lot of grunts, mostly), and then returning them to the divers.

I was actually in the dinghy, emptying a load of specimens, when I saw the head of a large green moray break the surface. It planed madly in circles about ten feet away, jaws open, showing its needle teeth. I wondered if it could snake its way over the side of the boat. Hard to tell how long it was, but what I could see of the body was as thick as my leg. It was at least twice the size of the eel that shredded my toe three weeks before. (As it turned out, I did lose the nail, but the doctor who sewed me up said another would probably grow back.)

I had all the collecting nets, but I was not about to get back into the water. The Eel Bay attack had been accompanied by many tales about the strength and ferocity of morays. Preston Sands had watched a big one chew the ash handle of a gaff to pieces after being boated; a friend of my father's had lost two fingers groping in a cavern for a wounded lobster. This eel could do much worse than shred a toe.

It had disappeared by the time my father surfaced next to the boat. "Gordy! What the bloody hell are you doing? Where are the nets?"

I told him about the eel.

"Where was it?"

I pointed.

"How big?"

I spread my arms as wide as they'd go.

"Good heavens! Stay out of the water and hand me the nets. I'll see if I can find it."

There was nothing more I could do, so I sat back and relaxed. The sun felt delicious. In the midday light there was no surface reflection off the water and each wavelet near the boat had its own particular shade of aquamarine. Dark coral heads dotted the sandy bay, and farther out the barrier reef showed yellowish brown before the deep blue of the drop-off.

Idly, I reached into a bucket of specimens to remove a two-inch, cigarette-shaped fish covered with tiny red dots and brown saddle-shaped bands on a white background. I had never seen anything like it. Could it be a new species?

I put the saddled fish gently back and pulled out a larger, olive, spade-shaped one with a white stripe on its forehead and black spots with pale borders all over its body like a trout. Another first.

Among the exciting new fish were a few fairy basslets, still jewel-like, even in death. Close up, they looked refined, delicate, sophisticated, advanced—more highly evolved than the other small fish. Oddly enough, I found myself thinking of them as specimens, and I didn't regret their dying. These fish might someday be very important. Exactly how, I wasn't sure at all.

Finally, my father surfaced with the green moray hanging heavily in his collecting net.

"You were right, he's a monster," he said. "A great addition to the collection. Put a little seawater in that empty bucket and pop him in."

I hoisted the heavy net and poured the eel into the bucket while my father sank back into the depths. Was it really dead? I poked it gingerly with the handle of a dip net; it didn't move. I ran a finger over its smooth, slick body, making a brownish trail in the green.

After five minutes of observation, the big eel was still motionless. Its body was as thick as my arm, if not my leg. Impossible to tell how long it was, curled up in the bucket, so I slid my hands under the middle of its body and lifted. No sign of life.

The dead eel hung limply down on each side of my hands. I stretched it out on the floorboards of the runabout, where it took up most of the space between the rear and middle seats. The only way to measure it, I decided, was against my own body: four feet, two inches long.

Lying next to the creature, my toes even with the tip of its tail, my eyes were level with its half-open jaws. The thin, translucent teeth filled the mouth in two roughly parallel rows and were, on the average, about half an inch long. The front teeth were half again longer than the rest, shaped a little like the canine teeth of a dog but more snaggly and irregular. On the side of the mouth nearest me I was able to count a total of twenty-seven, ranging from the long canines to the smaller, more regular cutting teeth in the rear. The entire eel was only four inches shorter than I was.

The year before, my father had built a little cinder block, tin-roofed laboratory on the harbor side of our property just east of the dock. There were six large fish tanks, a reference library, his growing volumes of field notes, barrels of specimens preserved in a formalin solution, and close-up photographic equipment. In a separate room were the aqualung tanks and gear, and a compressor.

Returning from Delaporte Point in the late afternoon, we went through our haul in this lab—125 specimens, including many that were unfamiliar even to my father: a spectacular new type of basslet with longitudinal black-and-red stripes on a yellow background, for example. Dr. Roberts's thin-lipped, almost expressionless mouth turned up at the corners.

"Charlie, this is very interesting. How many of these fish can we not positively ID?"

"A lot of the gobies and blennies. At least two of the eels. That odd-looking grouper, the candy-striped basslet. Many of the triplefins and the stargazers, and of course the red brotulid. Just off the top of my head."

Dr. Roberts nodded. "Off the top of *my* head, there might be room for a new handbook that includes all these discoveries. A good two-year project, at least. You'll need a bright-eyed, bushy-tailed postdoc to work with you."

I watched my father's face light up with a blush and a shy grin, like my own might have if he had said to me, "Oh, well *done*, Gordy." He'd just been awarded a project and an assistant . . . a real triumph. He shot a quick glance my way.

I might not have fully grasped the implications of what Dr. Roberts had said, but I'll always remember the glance.

Green Cay

April 1956

I WAS FLYING THROUGH WATER ALMOST as clear as air, the bottom about twenty-five feet down, the surface about twenty-five feet up. I was holding on to two wooden handles attached to the ends of a thin fiberglass foil, which was attached by a long line to a trolling dive boat. You controlled your depth by tilting the foil up or down. You banked by tilting it to the side. You breathed from the tank of compressed air strapped to your back.

Towers of coral rose forty feet from the sand bottom almost to the surface. In the hazy, sun-shafted distance, they were as imposing as the mesas in Monument Valley, blue instead of red. My father, towing to my right, swooped ten feet down, banked beneath me, and turned upside down to watch me, grinning around his mouthpiece. The hair on his chest rippled as the water rushed past. Without letting go of the handle, I made the circle of perfection with the thumb and forefinger of my right hand.

By this time I was about as expert as you could get with the aqualung given the fact that the sport was in its infancy and we used no buoyancy compensators, dive tables, computers, wet suits, or even depth gauges. "You can't get bent on one tank" was the rule of thumb. Which was true enough unless you went down again with another tank after too short a surface interval. We lived charmed lives, though, and that never happened.

The water off Green Cay, the scrubby, isolated little island a few miles north of Eel Bay where we were towing, was fifty feet deep, deeper than I'd ever dived before. Looking down could make a person vertiginous: you had a strong feeling that if the boat stopped and your forward movement ceased, you'd fall like a kite when the wind dies. God only knew what giant fish lived out here. I kept checking behind me for the twelve-foot barracuda that might think I was a trolling bait.

We were looking for the perfect coral head to collect on: steep and pillar-like rather than squat and spread out, so the cloud of poison would rise up through it and the poisoned fish would sink to the sand bottom around it instead of disappearing in the caves.

Jim Böhlke, the bright-eyed, bushy-tailed postdoc whom Dr. Roberts had hired right out of Stanford to work with my father, was watching us from the boat and telling the captain which way to turn when we surfaced to point. This was his second day in the Bahamas, and his skin was crimson with a bad sunburn.

I could see my father suddenly sheer off away from me, banking like a bird, saw him point at a dim coral tower, point his thumb to the surface, and release his foil. I understood that I was supposed to go up, tell them he'd found it, and guide them there.

With the boat finally anchored in the sand next to the head, I could see how perfect it was. Erupting from the bottom like a twisted pillar of fire, it was wider at the top than at the base, the living coral there flaming pale yellow into the blue depths around it, clouds of blue chromis, yellow wrasse, silvery bogas, and striped juvenile parrotfish swirling above it like smoke. Free-swimming silver jacks sailed through the blue, and sure enough, a smallish barracuda eyed me from a distance. It might have been larger than it looked, but with my trusty aqualung I felt more like an equal.

Working my way slowly and vertiginously (*"Watch out, don't fall!"*) down around the promontories, I could see the whole structure was fissured with a large crevice, begging for investigation. Fifty feet down, a flat sand bottom stretched off into the blue. From there the surface was barely visible. My father and Jim Böhlke hung at different levels on the tower, like spacemen, barely moving, their breathing punctuated by rising clouds of light-mirroring shimmying half-domes.

During a lunch break on the warm deck of the captain's converted shrimp boat, we discussed our plan of attack. Jim and my father would take their pillowcases of rotenone down to the base of the reef, on opposite sides, inundate the deep caves, and work their way up until the sacks were empty. When the first poisoned fish started to appear fifteen minutes later, my sister, Susie, and I would jump in and start netting them with long-bodied scoop nets, me with my aqualung while she patrolled the surface with snorkel gear.

She was nine years old at this point, blond, freckled, serious, and determined. Very easy and fun to tease.

"I bet I get more fish than you do," I told her after Jim and my father had gone over the side with the rotenone.

"No fair. I can't dive."

"Sure you can. I've seen you. You're good at it."

"Yeah, but you have a tank."

"With a tank you can't move as fast. Plus there's more fish on the surface anyway. Bet one of my butterflies against one of your cowries. Winner gets to choose." Our father had lent her the cowries from his own collection, but I was pretty sure that if I got my hands on one I could keep it. I had collected the butterflies myself.

She thought about it for a minute and finally agreed, as I knew she was going to. I also knew which cowrie I was going to select, a beautiful, bubble-shaped dark brown one with black spots from somewhere in the Pacific.

In the water, poisoned fish were darting everywhere. My father was working his way cautiously into the deep fissure. From a ledge about twenty feet down, I could see Susie on the surface, swimming around at full speed to gather them up. The creole wrasse and chromis had all gone immediately up there, so she had an advantage. The long belly of her net was already about a quarter full.

At least for the moment. Fish from the interior of the reef were more numerous and in time would sink to the bottom. Or so I figured as I worked my way down.

Already, in fact, the sand flats at the base of the coral head were littered with candy-striped basslets, fire-engine-red flamefish, midnight-black gobies with one yellow lengthwise stripe, and dark green gobies with fluorescent light green bands, white heads, and red eyebrows.

We were now sure that rotenone did not have lasting effects. Over the years we had gone back many times to the same coral head off Delaporte Point where we'd made our first collection, to observe the fish life undepleted. So I didn't feel bad about the carnage. Finding the little fish and collecting them was like finding Fabergé jeweled eggs in an Easter hunt, but much, much better. They were more beautiful than Fabergé eggs, and any one of them could be a new species.

You could almost lose yourself in the deep dark caves at the base of the coral head, and underwater flashlights had yet to be invented. I'd penetrate to the edge of visibility, stirring up the bodies of tiny fish and eels with my hands, then scooping them in.

A shaft of light through the reef lit up the very end of one of these caves, and lying in it I could see a bright red squirrelfish. The squirrelfish had a dark blotch at the base of its tail and another at the end of its spiny dorsal fin and looked different from any squirrelfish I'd ever seen before. But the cave narrowed down, and it would be hard to reach.

If anyone could get it, I could, since I was by far the smallest. As I slowly worked my way into the cave, my chest scraping the sand bottom, my tank occasionally clanging against the rock ceiling, fifty feet of water pushing down on me, my eye was on the prize and that was all I thought about. You might have called it foolhardy to the point of insanity, but in the back of my mind was an abiding faith that no harm could come to me because my father was around.

Even so, to bag the fish I had to push and stretch, pawing gently with my net, agonized that the prize would drift out of reach in the eddies.

When the fish was safely inside the belly of the net, I realized that I could hardly move. There was no question of turning around. . . . I'd have to back out, pushing myself with my hands and feeling my way with my feet. Inch by inch, like a lobster. The bottom of my tank, riding high over my back, caught on a ledge.

To clear it, I'd have to reach around and pull it down, but in the narrow space I couldn't get an arm loose. I was puffing and panting with adrenaline and effort, using up air at a tremendous rate. Now it seemed harder to breathe. Was my little tank running low?

Before I had time really to wonder about how things would turn out, I felt a hand on my ankle. I relaxed and went limp as another hand reached along my body to clear the tank. Then, as if by magic, my body was sliding backwards over the sand. In no time at all, I was sitting on the flat sandy bottom surrounding the coral head, nodding at my father, pointing at my net, then making the circle of well-being and triumph with my thumb and forefinger.

He jerked his thumb upward, and we both rose slowly, side by side, in a cloud of mercurial bubbles. He was on his second tank, so he stopped ten feet below the surface to decompress while I continued up, noticing my air was harder to suck out than ever.

Jim Böhlke was aboard the boat, and I handed him my net.

"New squirrelfish, I think," I called, shrugging out of my harness, tying it off to the tender line, and hurrying up the boarding ladder.

He dumped my catch onto the sorting table and separated out the squirrelfish, turning it over and over in his hands. Finally he gave me a wide feel-good smile: "Yup. I think we've got something here."

"What is it?"

He pointed out the two dark blotches I'd noticed. "I think it's a saddle squirrelfish, *Adioryx poco*. Also, look, its got thirteen rays in the dorsal fin while the reef squirrelfish, *coruscus*, has twelve. Only four pocos have ever been collected, as far as I know. This would be the first one from the Bahamas. Where was it?"

My father had climbed aboard by that time, and heard the question. *"Duh-hunh!* You might well ask." His voice sounded odd.

Jim Böhlke's eyebrows went up. "Uh-oh. Did he pull a Randall?" I'd heard the stories about Jack Randall, a legendary ichthyologist who had collected all over the world and had described far more species than any of his competitors. He'd gotten trapped in a cave similar to mine, trying to net an interesting specimen.

"Jack jettisoned his gear and got out alive." My father nodded his head in my direction. Was it my imagination or did he look proud? "*Gordon* would probably still be there if I hadn't happened by." He pulled on the line attached to my tank, hoisted it into the boat, turned off the air, and unscrewed the regulator. Then he put on an air gauge and reopened the valve. The needle barely moved. "Look at that."

The captain padded over, looked at it himself, and whistled. "That's cutting it close."

A long silence. "I'm sorry," I said finally. "I just . . ."

Still no one spoke.

"Don't tell Mummy. Please?"

Jim Böhlke stifled a laugh. Then he said: "There speaks a true scientist." And yes, my father did look proud, cocking his head with a little grin and twinkling eyes. Ah yes! I could bask for a minute or two in the heavenly warm sun of his approval.

Susie climbed aboard elatedly, her net brimming. My father emptied it onto the sorting table. "Well *done*. A veritable cornucopia."

"How many did Gordon get?"

He did not look up from the pile of fishes. "You were both fantastic."

"But aren't you going to count them?"

"Of course we are, when we get back to the lab."

"But how do you know who got them?"

"Oh, we don't keep track of that. Unless they're very unusual. Then we'll make a note about who was the collector."

Susie sensed something was up. Somehow she was being short-changed. I'd come out ahead yet again in a way she didn't fully understand. All she knew was that something precious and very hard to get was being withheld from her and given to me.

I saw her looking at me then in a way that was going to become very familiar.

Andros

July 1961

THE WEATHER WAS HOT AND squally on the first leg of our collecting expedition across the Tongue of the Ocean, the 6,000-foot-deep, twenty-mile-wide underwater canyon that separates New Providence from Andros Island. As the *Sea Pal* began to roll in the steep purple swells, I gripped a stanchion, kept the wind in my face, and tried to take the growing queasiness like a man . . . even though, at sixteen, I was still squeaky-voiced and smaller than all of my classmates at boarding school.

I was being included on this trip, in fact, as a reward for having squeaked through yet another year at this school, chosen by my father as the closest thing to Eton he could find in America.

He'd hated Eton and hadn't done well there, but still wore his blue-striped school tie with pride, attended Old Etonian gatherings in New York, and kept an inscribed china pitcher that played the *Eton Boating Song* in a place of honor on the mantel. He must have hoped I could succeed in this arena where he had failed. It was the same unfruitful pattern my mother followed with Susie: a tony boarding school she hated, cute little outfits she loathed, subdeb parties and dancing classes where she felt awkward and out of place.

The strange thing was, I never felt much pressure from him as my years at this school slipped idly by. Not getting thrown out was all he seemed to require. Teachers were constantly telling him that I could do better, that I was a classic underachiever, and I was only to read in his diaries after he died the cryptic line: "I wonder if Gordy is as smart as his teachers keep saying he is?" Was that his way of exonerating me?

It was as if my mind had gone on strike and agreed only to do the bare minimum. Of course, the experience of slowly turning into a shrimp among your peers leaves lifelong scars. But the worst thing of all was being exiled from the paradise of Nassau to the Siberia of Concord, New Hampshire, for no reason that I could understand. Sending me there against my will was the beginning of a long road of disaffection.

But for now, fuck my classmates. I was on this boat, having this adventure, and they weren't. I'd face down the queasiness, no problem. My father and the captain were chatting at the wheel, their backs to me. Dr. Roberts and his son, Raddy, were nowhere in sight. Jim Böhlke was braced in a deck chair trying to read a textbook.

A foot-long flying fish launched itself out of a wave top, bright silver with vibrant, transparent wings against the purple ocean, glided over several crests, banked, and splashed down. Jim had been watching.

"You see the size of him? Mirrorwing. Pretty rare."

I opened my mouth and closed it again, like an unwell fish.

"And delicious." He smacked his lips. "Sauté 'em in butter, they're the best fish in the world."

My stomach flip-flopped. I passed a hand over my damp forehead.

He looked at me more closely. "Feeling okay?"

"Oh yeah . . ."

"Pretty rough out here. If it starts to get to you, try eating a slice of bread or a cracker. That'll help."

I was ready to try anything, so after a discreet interval, I staggered down into the galley. Through a cabin door I could see Raddy Roberts lying in his bunk, face to the wall. I grabbed some bread from the bread box and just made it back up in time to clutch the leeward rail, bend over, and blow my breakfast.

Had anyone noticed? Jim seemed to be buried in his textbook. My father was still deep in conversation with the *Sea Pal*'s captain, and Dr. Roberts was still nowhere in view. I was standing there chewing the bread, watching the waves, and trying to make friends with my stomach, when I felt my father's hand on my shoulder.

"All right, old thing? I feel a trifle peaked myself."

I nodded and grinned as well as I could.

"You know, Nelson used to turn positively *chartreuse* at the beginning of a voyage. He was that way all his life."

"Nelson?"

"Lord Nelson, the hero of Trafalgar."

"Oh."

"Yes. But after a day or two on the briny, he'd be chipper as a cricket again."

A day or two?

I was able to negotiate a truce with my stomach for the next couple of hours as the distant rise of barren rock that was High Cay, marking a passage through Andros Island's 140-mile-long barrier reef, got gradually closer. Then, suddenly, the bottomless purple ocean changed to light blue and we were in calm water.

"This is more like it," Jim Böhlke grinned. "Am I right?"

When the two Robertses finally lurched up from below, my father raised his eyebrows at me and comically pulled down the corners of his mouth.

Almost noon. The *Sea Pal* chugged south over the shallow flats inside the reef, picking its way around easily visible coral heads—dark brown against the white sand bottom twelve feet down. Half a mile to the west, the main island showed as a narrow white beach backed by casuarinas and palms, with absolutely no sign of life. Rorschach summer clouds slowly changed from one shape to another. I munched on a ham sandwich, the queasiness a distant memory, enjoying the fact that Raddy Roberts (two years older) had palely refused his.

I had never seen anything like this vast, spectacular wilderness, but for some reason I felt at home here. The drifting, changing clouds and their shadows made the scale of things seem even grander than it was. The water here looked twice as clear as Nassau's: the slightest variation in depth registered a corresponding nuance in color. And we seemed to be the only living souls within hundreds of miles. (This was not quite true, but it was pretty to think so. In any case, Andros's 2,300 square miles of mangrove swamp and hardwood thicket includes one of the largest tracts of unexplored land in the Western Hemisphere.)

The water boiled strangely up ahead over a small patch reef. As we circled, we could see this was caused by upwelling tidal currents from a cave in the reef's center, a blue hole, somehow connected underground to the Tongue of the Ocean. Jim thought collecting here might be productive, so we dropped anchor and hopped overboard to reconnoiter with masks and snorkels.

A curtain of foot-long grunts, bright yellow with electric blue stripes, hundreds and hundreds of them, more grunts than I'd ever seen together in one place, hung in front of the reef flashing their luminous eye-whites while they watched us. As my father and I approached, they flowed aside to reveal thick stands of elkhorn coral reaching so close to the surface we had a hard time squeezing over.

On the downslope into the cave's narrow mouth, the strong upwelling current was cold, almost icy. A few big dog snappers loitered close to the lip, disappearing inside as we flapped our way down toward them. Hanging on to the lip, we could see them holding at a bend in the fifteen-foot-wide shaft another twenty feet down.

I had some breath left and was proud that I could free-dive to more than forty feet, enough to get to the bend and see what was beyond it. I was screwing up my courage, but my father beat me to it . . . receding into dimness as he finned strongly against the current down the shaft, pausing for a slow count of five at the bend, then letting the current carry him back to the surface next to me.

He was grinning. "Very spooky. A large cavern leading down into the dark, with no bottom. Very spooky indeed."

"Can we dive down there with aqualungs?"

The grin widened. "This one looks risky. A lusca could be lying in wait."

"A *lusca*?"

"They live in some of these holes, you know. Waiting to nab the foolhardy."

"Oh, come *on*, Daddy." He had to be joking. "What do they look like?"

"Ah." He laid his right forefinger along the side of his nose. "Unfortunately, no one has ever returned to say."

My father loved the idea of monsters, elves, creatures from outer space, unexplained supernatural forces, and beings of any kind. He read the *National Enquirer* religiously, and at the most staid and soberest dinner parties was always ready with a juicy tidbit: "Did you know they found a werewolf in Maine the other day? Looked very like a dog, but with the face of a vampire bat. There was a photograph of it in the paper. Most amazing."

He'd already filled me in on another strange creature indigenous to Andros: the chickcharney, said to be indirectly responsible for World War II. Shaped like a large bird, with oversize red eyes and three toes on each foot, the chickcharney casts no shadow and builds its nest only in certain pine trees that grow densely enough so that three can be drawn together in a tripod to support the nest. Chickcharnies know miraculous cures and trained the famous Seminole medicine man Billy Bowlegs, but if you mistreat or annoy them they put a "sign" on you that blights you for life.

Neville Chamberlain, prime minister of England from 1937 to 1940, found this out the hard way according to legend. Running his father's sisal plantation on Andros in the 1890s, he ignored local counsel and ordered a group of chickcharney trees cut down to clear new ground. In a few years, his plantation failed (along with the entire Bahamian sisal industry), and Chamberlain returned to England to enter politics. As prime minster, his policy of appeasement toward the Nazis emboldened them to invade Poland and start the war. At this point in the story my father would nod his head wisely, eyes twinkling. "Let that be a lesson to you, my lad."

We pulled into South Bight at about four p.m., July 10, and anchored in the middle of the channel near another boiling ocean hole. The palms and houses of Lisbon Creek were visible on the north shore, and on trestles under the palms we could see men at work on the hull of a new sailing vessel.

My father took Raddy and me in the *High Hat* to the south side of the bight to look for tarpon. On the way over, I spotted the dark triangular shape of a large spotted eagle ray swimming slowly along just above the bottom, looking for mollusks to eat.

We had a harpoon-like pole, powered by surgical rubber tubing attached to the back end: you looped the tubing around your hand, slid your hand up the pole to stretch it tight, then gripped. To fire, you simply let the pole slide freely through your hand. We had a fishing rod with fifteen-pound test monofilament line. It was my idea to attach the line to the harpoon, harpoon the ray, and play it like a large fish until we could land it.

My father's eyes lit up. "Excellent plan, Gordy. We'll give it a try." He tied on the line and handed me the harpoon. "Get up in the bow. I'll try to work in close enough for a shot."

Closer and closer. The big ray flapped powerfully on like some huge underwater bird, paying us no mind. I could see that some of the white spots on its back and wings were actually circles. Its cow-like snout was decorated with white stripes. Its thin black tail stretched six feet behind it, with two long spines at the base. "A stab by the tail spine is described as excruciatingly painful," reads the text in *Fishes of the Bahamas.* "In some cases the loss of an appendage or even death has resulted from such a wound."

The moment had come. After I loosed the harpoon through the clear water, I could see the point of the spear enter the black hide just to the right of the spinal ridge. A quick shudder as from an electric shock, then an explosion.

At first it seemed we were going to lose all this line, as the ray headed for the horizon as invincibly as an F-111 fighter, but Raddy turned the boat after it and kept the distance manageable. We took turns playing it until we were all soaked with sweat and our arms drooped like wet noodles. Almost an hour later we horsed it into the shallows near Lisbon Creek, gaffed it, and dragged it up on the sand. Staying clear of the spines was easy enough—you would have had to step or fall on the tail's base to get stabbed.

The big ray lay there, spread out like a black-and-white leopard-skin rug, motionless and otherworldly, the largest creature I'd ever speared and probably ever would.

"Well done, boys," my father said. "An epic battle."

Thanks to him, I'd cut my teeth on *Moby-Dick*, and today we'd actually gotten our quarry instead of being destroyed by it. He extracted the spear and touched the ray's tail-spines with the point. Their mucus coating glistened in the sun. Under the mucus you could see saw-like serrations. "They go in quite easily," my father explained. "But pulling them out against the barbs lacerates the victim's flesh and releases the poison."

Raddy wanted to know what kind of poison it was. Like a rattlesnake's?

"*Much* more painful. A snakebite hurts very little to start with. This is like being stung by a giant wasp."

"Is it worse than a stonefish?" I asked.

"Well, of course, nothing's worse than the dreaded stonefish," my father answered with relish. One of his favorite books was Halstead's *Poisonous and Venemous Marine Animals of the World*, with its graphic description of stonefish poisoning. He'd read it to me once, and I'd gone back to the text many times in fascination: "The pain, which is generally described as instantaneous, intense, sharp, or burning, radiates within a few minutes from the wound site, involving the entire leg, groin, and abdomen, or if in the upper extremities, the axilla, shoulder, neck, and head. The pain might become so severe that the victim thrashes about, rolling on the ground and screaming in agony, and may lose consciousness." Luckily, stonefish are found only in the Pacific. A local relative, the Bahamian scorpionfish, is pain-wise not even close.

Raddy took the *High Hat* back to the *Sea Pal* and picked up the others. We photographed the creature (to be reproduced exactly in *Fishes of the Bahamas*, right down to its slightly shortened tail, probably as the result of injury), weighed it with a game-fish scale (150 lbs), and took measurements: 170 cm from wing tip to wing tip, snout tip to dorsal origin 107 cm, eye 19 mm in diameter, et cetera. Inside its stomach were the remains of oysters and clams, almost intact and without the presence of shell fragments, showing how skillfully the ray had separated those out after crushing the shellfish with its cement-like dental plates—the projecting lower plate being used like a spade to dig up the mollusks. We took tissue samples for preservation.

We finished just before dusk. Then we towed the remains out to the *Sea Pal* and put them on a foot-long shark hook attached to the boat with a three-eighths-inch nylon line. Jim Böhlke smacked his lips. "Nothing tastes as good to your average shark as a nice fat eagle ray."

In the moonlight after dinner, with a rising tide perfect for shark activity, we sat in the cockpit watching and waiting. Around nine p.m. we began to pick out the dark shapes circling the carcass, and then the line went taut.

The line was cleated to a sturdy wooden docking post in the stern of the boat. The shark at the other end pulled the stern around against the current and kept it there for half an hour before tiring. Finally, my father and Jim Böhlke hauled the five-foot creature in close enough for the captain to gaff. Then they got a rope around its tail, hauled it half out of the water and cleated it off, to be weighed, measured, and dissected on the beach in the morning. Sweaty and exhausted, Jim pronounced it a bull shark (*Carcharhinus leucas*). Half an hour later, we caught a six-foot tiger (*Galeocerdo cuvier*) the same way.

Both species are considered dangerous to swimmers. About the tiger specimen, *Fishes of the Bahamas* observes: "The individual pictured above was taken on a foot-long shark hook baited with an entire eagle ray, five and one half feet wide; at this size, the eagle ray "wings" are thick and extremely tough but before being hooked, the tiger had neatly and effortlessly bitten about half a dozen large and regular chunks from the ray." (We saw this after dissection the next day.)

The ray was only half eaten, the sharks were on the move, the night was young, and, tired as we were, at least some of us were on a shark-fishing high. We put the bait back down, but after an hour or so nothing had happened. The first flood of tide was now over, my father estimated.

"My bunk's going to feel mighty good," Jim Böhlke said, and soon I was alone in the cockpit, watching luminous clouds sail past the moon, still too excited to think of sleep.

Ten years earlier, I had destroyed Winston, the little octopus, because he threatened the peaceful world of the aquarium. You could imagine creatures like sharks, barracudas, morays, and eagle rays as threats themselves, but I was beginning to realize that they were more than that. Much more. Without them, in fact, the underwater world would only be beautiful.

At around ten thirty, I happened to glance over the *Sea Pal*'s cockpit coaming to see a huge shape circling under the boat in the moonlight. I yelled, and all hands emerged blearily from their bunks.

"Hammerhead," Jim Böhlke pronounced. "Over ten feet!"

Soon enough, the line went taut with a steady inexorable pressure that slowly pulled down the fifty-foot *Sea Pal*'s stern until the surface of the water was only a foot below the deck. The shark was not fighting, not panicking. It was just . . . moving off.

"Holy Mother of God," I heard the captain mutter. The nylon line seemed to hum in the air like a giant cello string. Droplets of water flew from its vibrations as it sliced the surface, and the stern of the *Sea Pal* was drawn lower and lower.

At the last minute the boat suddenly lurched back up. My father pulled in the slack line to find the big tempered-steel shark hook completely straightened out.

"Will you look at that?" he said in wonder, respect, and undisguised pleasure. "Good lord what a brute!"

So, a few days later, on an offshore head in about fifteen feet of water, it seemed totally appropriate for our final collection to be interrupted by a large hammerhead shark sweeping in to devour the dying fish.

Dr. Roberts, Raddy, and I were already back in the *Sea Pal* when the badly peeling Jim Böhlke (dressed in a diving costume of white Oxford shirt and khaki pants to shield his sunburn) clambered over the side and told us about it: "I just suddenly felt like I should take a look behind me, so I pulled my head out of the grotto and there she was. Hello! She was so close I had to look from one eye to the other. Should I drop my fish? No way. She was a well-fed shark by that time, anyway." He laughed. "Where's Charlie?"

We all looked at each other. "I guess he's still over there." I pointed to the far side of the head. "At least that's where he was when I left."

"I guess somebody should let him know," Jim said.

A moment passed, a bit uncomfortably.

"Okay, I will." I started putting my gear back on. I had never been in the water with a large shark, but how could it be as terrifying as a large barracuda?

I could see the two men weighing the pros and cons: I was sixteen, old enough to take a risk to help my father. But if I were actually attacked, they'd be responsible.

Jim Böhlke, who'd been present at my rescue off Green Cay a few years earlier, began to nod. Dr. Roberts finally acquiesced.

"Go ahead, Gordy. But please be careful."

In the water, swimming as fast as I could without appearing to be frightened (fish can sense fear as well as they can sense aggression), everything seemed brighter, clearer, and generally more vivid than ever before. There was a new dimension, not exactly an unpleasant one but not one that you'd seek out on purpose. Or was it? Stan Waterman, my father's best diving buddy, loved to talk about his shark adventures. People said he actually loved the sharks themselves.

Finally, I saw my father's bubbles rising on the other side of a coral promontory. His body was in a grotto, like mine had been when he'd rescued me. I could see only his fins, slowly waving to keep him in position. I jackknifed, dove fifteen feet to the bottom, and gripped his ankle exactly as he'd gripped mine. When his head appeared, I gestured wildly around and jerked my thumb upward. I was now hoping that the shark would show up.

"Hammerhead!" I told him on the surface. "Jim Böhlke saw it! Everybody's in the boat!"

"Gawd. How big?"

"Jim said he had to look from one eye to the other."

"Damn! The brute's going to eat all our fish."

Perversely, there was no sign of the shark on the swim back either, and I was able to formulate a crude law of nature: the more you want to see a particular wild creature, the less likely it is to appear.

"Gordy wanted very much to be the one to tell you," Dr. Roberts told my father when we were safely aboard.

He raised his eyebrows and gave me a courtly nod.

"Many thanks, old fellow."

I'd never learn if he had made the connection I'd wanted him to make: he'd rescued me, and now I'd rescued him. We were even.

For most of the Andros trip I felt like a wide-eyed child again, drinking it all in and savoring every drop. But then, we all felt like that, even Dr. Roberts. I could see it in his face.

We were doing serious things and having the time of our lives. I suddenly thought I saw the connection: it all added up to science. Science was the passport into this magic realm, where important discoveries could be made in exotic places by people who knew their stuff but still knew how to enjoy themselves. Raddy Roberts was planning to be a scientist like his father. I could too, though it would mean going one up on my own father, who after all was just an amateur. Come to think of it, besting the man who had exiled me from paradise might be one of the better parts. It might show him he'd made a mistake.

Andros Island was the catalyst. Without that wild, remote backdrop, I don't think I would have seen the connection. In one way or another, I've spent a good part of my life trying to get back there. I've lived in Mexican deserts and Southeast Asian rain forests, and sailed halfway across the Pacific, but nothing measured up to my teenage memories of Andros on that trip: a solitary boatload of happy scientists on an unexplored azure sea.

Entr' Acte

1964

"THANK GOD WE WEREN'T THERE," my father wrote in his diary after I was thrown out of Cornell University's undergraduate wildlife conservation program. He and my mother were traveling in Europe when they got the news, and they didn't come back. I was glad. They'd just be pissed off. And anyway, my aunt had agreed to take me in.

I'd blown it royally, showing up for spring break at our house in Nassau with a horde of brothers from my new fraternity, only to find it rented. In high *Animal House*–style, we'd terrorized the tenants and the neighbors, crashing in beach shacks, on park benches, and on buses, finally returning to college a week late to face expulsion for the lot of us.

In my case, it was more than running wild. I was grinding my heel in all the good things that had come my way on Hog Island, my home, dragging my parents' respectable name through the mire, spurning my studies, creating an antihero persona for myself that would take years to live down.

Maybe, if I'd liked my studies, things would have been different. I'd picked the Cornell program myself with specific aims in mind. But I couldn't see the connection between my grueling zoology and botany courses and the joy of being in and around the ocean. The trail of footsteps I'd been trying to follow petered out in untracked sand. My father, in his laissez-faire English way, hadn't tried to make them clearer.

"*Thank God we weren't there.*" From my jaundiced point of view as an exile, it seemed quite possible that my father hadn't *really* been there for me since that brilliant fall day when he'd dropped me off at the Lower School dorm and driven his silver-and-midnight blue Bentley down the elm-lined drive toward the campus gates and out of sight.

He should never have sent me away from the place I loved and was thriving in. I couldn't forgive him for that. And I don't think he ever really forgave my smoldering apathy at the school he picked for me, and the public way it finally caught fire on Hog Island.

My aunt Louise, first cousin to my mother, lived in New York with her youngest daughter. Her husband had died a few years earlier of a heart attack, and her two older children were grown. She gave me bed and board but no allowance, because with money I'd have freedom and that was not what she wanted. To get to the roots of my academic flameout, she had me going to a psychiatrist an hour a day—she was a great believer in shrinks. Meanwhile, I was supposed to be looking for a job.

My cousin was sixteen and went to a fancy girls' school around the corner on East 83rd Street, across from Carl Schurz Park and Gracie Mansion. She was the opposite of me, but exactly. She worked very hard, got straight As, and never got into any kind of trouble. Her friends were all respectable (although not that numerous). She was good-looking enough so she always got the benefit of the doubt, not that she needed it.

As you might imagine, my aunt was very careful about the kind of contact my cousin and I were allowed to have. Tennis Tuesdays and Thursdays and a rare chaperoned movie were about it. We played in the late afternoon after her lesson at a club on Central Park West (dovetailing nicely with my hour at the shrink's, whose office was nearby). Weekends, she often went to visit girlfriends. I had no friends in New York and no money to go out on my own, so my social life was zero. My aunt decided it would be therapeutic for me to have a guitar. I would practice blues riffs for hours then immerse myself in Kerouac's *On the Road*. I felt I could go in any direction, like Sal Paradise, hitchhiking out west to make it on my own. All I needed was a sign.

Most evenings, the three of us would have dinner together. The format was closely organized by my aunt into three parts for each participant. Review of day. Selection and exposition of day's theme. General discussion. My aunt would cook; my cousin and I would clean up. Sometimes during the dishwashing we'd play little games (Squeeze the Soap), and our hands and shoulders might touch. Then it was time for her homework and my book.

The tennis club we played in had a green court, green walls, and dim illumination. Her white tennis outfit (Bermuda shorts and a sleeveless blouse) glowed with its own light. The whites of her eyes were so sharply defined, her irises (which were actually light brown) looked black as they snapped around the court. When she worked up a sweat, I could clearly see the outline of her bra. Her bras were stiff and substantial without undue emphasis, bought no doubt under the guidance of my aunt.

Sometimes, when she won a close point, she'd grin saucily across the net and I'd hit balls at her with longing. We'd take the crosstown bus home, holding our rackets between our knees. Our knees would touch on the bumps. The smell of her scented soap and perspiration would surround us. I looked up a word in the dictionary for the first time in years: "*Longing: an eager desire esp. for something remote or unattainable.*" It gave me new respect for Noah Webster.

"Do you think about girls at all?" the lady psychiatrist asked me one afternoon before tennis. I froze. My heart slammed. I was convinced she could read my mind and the next step would be to inform her friend, my aunt. I would be out on the street.

I thought about my cousin all the time. After I had finished with her, her white tennis outfit would be torn and stained, her lips would be swollen, her face bruised, her neck marked with teeth, her beauty rumpled like a used bed. She would be beneath me. Below even the level I was on. And she would love it.

"What are you thinking about now?" the lady psychiatrist asked.

"Food."

"Any particular kind of food?"

"Just food. I'm starving." I thought as hard as I could. "Maybe pork and beans. Something filling."

For dinner, my cousin dressed in green corduroy slacks and a heather Shetland cardigan buttoned up to the neck. Her thick hair was damp from her shower and pushed back behind her ears. In an hour or so it would be back to its full chestnut waviness. The apartment windows were open for the first time since I moved in. It was a warm spring evening. A little breeze off the river, and you could hear the tugs.

"I've got a job interview tomorrow," I announced during the review-of-day period. "Lab assistant. I was majoring in biology, after all. Maybe they'll take me."

"Good for you! Where is it?"

"Morris Avenue, Brooklyn. They told me to take the F train."

After a short silence, my aunt shook her head. "That's much too far. You should be looking for something here in Manhattan, as I said in the beginning."

"But they said it was only twenty minutes."

"Take my word, it'll be very inconvenient."

"I've been answering ads for weeks. These people are the only ones who even asked for an interview."

"You've been very industrious, and I hope you'll continue. But Brooklyn is out of the question."

Brooklyn! The idea of the place began to take on an obscure weight and meaning. "Do you know anybody that lives in Brooklyn?" I asked after a while.

"Certainly."

"But I mean, friends?"

"No."

"A lot of people live there, don't they?"

"Of course."

"What kind of people?"

She looked at me carefully. "People who are trying their best to get out, to be able to afford a better place. But it's hard for them because they're from poor countries. Or else they're Negroes." A tentative smile. "Look at it this way: they're all *desperately* trying to get where you already are."

She got up and went into the kitchen where she was broiling chops. My cousin and I stared at each other over the soup dishes. "I think Danny Kaye comes from Brooklyn," she said.

I laughed, picked up my fork and drummed it loudly on my dark green–bordered Currier & Ives place mat "I don't."

"Damn," we heard my aunt saying above the sound of sizzling chops. "Will someone come and help with these?"

My cousin started to move, but I was too fast for her. I was on my feet and through the kitchen door before she was out of her chair. Beside the chops, which were a little burned, my aunt had green beans and butternut squash. I told her to go in and sit down while I put the food into serving dishes and brought it out.

The food looked rich and filling and smelled delicious. I was so hungry I was literally drooling: a silver line of saliva suddenly connected the corner of my mouth to the saucepan of squash. I put the chops on a silver serving platter, poured the gravy over them, and carried them out.

Instead of putting them on the table as usual, I served them from the left with a little flourish, like a waiter at "21." My aunt and cousin smiled with appreciation and cocked their heads.

Standing beside my cousin, looking down at the straight, clean part in her chestnut hair, I felt my throat tighten and my chest catch. Before I knew what I was doing, I had dumped the platter down the front of her sweater into her lap.

Of course I was forgiven. The platter had slipped. I was just as shocked as they were. We stared at the mess in silence without moving, and then my cousin looked up at me. She looked at me from under her eyebrows, her head cocked slightly to the side. There was a smile somewhere in her face, just a hint . . . no, nothing . . . you couldn't be sure. No matter how many times I played it over, I still couldn't be sure.

To celebrate my promotion from office boy to canvasser, my aunt had a surprise. Dinner for the three of us at the Rainbow Room in Rockefeller Center. I would go to my shrink, play tennis with my cousin as usual, then we would take a cab down to the restaurant and meet my aunt. She announced all this with a fairly tentative look, as if she knew what had been going on inside my head.

Actually, my cousin and I had not had much contact with each other for a while. Final exams were coming up for her. I'd been working at Republican campaign headquarters for the Silk Stocking congressional district, a volunteer job my aunt had arranged through friends to get me out of the apartment. Nights, I was now reading Hubert Selby Jr.'s *Last Exit to Brooklyn*. It was the most grisly book I'd ever read, and I couldn't put it down.

My cousin was hitting balls against the backboard when I arrived at the club. She was wearing something different, a short white dress with a vee neck emphasized by a sky blue stripe. My aunt would make her take it back the instant she laid eyes on it. My cousin waited for me, a white shadow, blushing and fidgeting as I walked across the green.

"Do you like my new outfit?"

I played as hard as I could and beat her 6–0 in three straight sets, her ruffled panties visible on low returns.

Showered and dressed for dinner, we walked slowly south on Central Park West, watching for a cab. The trees in the park, with their new leaves, were in shadow, and the setting sun burned on the buildings across the park along Fifth Avenue. My cousin was wearing her best spring dress, the one she wore for company: a dark blue, full-skirted taffeta with a scooped neckline discreetly showing her décolletage. For the first time I knew of, she had low heels instead of flats.

She did not talk much; I had never beaten her so badly. Occasionally, I would stop walking and pretend to look at something in one of the apartment house lobbies. She would stop too, and try to figure out what interested me.

We were walking past the Dakota, which my aunt would have looked down on as the abode of "theatre people." The huge windows on the first floor were not too far above street level. A gray-haired man was standing in one of them watching the sunset. He winked when he saw us. Right next to the Dakota were green illuminated balls marking the subway. I started down the stairs.

I expected my cousin to stop, but she kept up with me, the sound of her heels echoing in a grown-up way.

"We're never going to get a cab at this hour," I explained. "This'll be much faster."

She just nodded.

I paid both fares and handed her the token. She followed me through the turnstile.

"Have you ever been on a subway before?"

She shook her head. "Mom says they're dangerous."

There was no one else on the platform. We walked up and down, listening to the sound of her shoes. I planned on taking the D train when it came by: the express to Brooklyn.

We sat on a bench to wait. Now the only sound was a faint rush of traffic from above. Out of the corner of my eye, I could see my cousin's blue-skirted knees. She had put on perfume for the occasion, faint and tasteful, like lilies of the valley. I figured the D train would make just a stop or two more in Manhattan, and then we were gone.

In Brooklyn something would happen to us, something drastic, so drastic I couldn't even begin to imagine it. And things would never be the same.

"I got you a present," my cousin said.

I didn't answer. I was afraid my voice would come out as a growl.

"Well?" she said. "Aren't you going to say thank you?"

"Thank you."

"I was going to wait until dinner, but I guess I could give it to you now." She was looking at me from under a lock of still-damp hair. I must have looked dubious, because she made a pretty moue, shook her head, and added, "If you *want*."

I tried to smile, and nodded. "Of course I want."

"But first," she said, the lock of hair falling farther over her eye, "you have to tell me: do you want your kiss before or after?"

From far off down the track came a breath of warm, stale air and the faint squeal of iron on iron. I could imagine a ghostly line of deserted lit-up cars orbiting through the dark tunnel.

I took a deep breath. "Before."

She arched her neck and whispered something in my ear. Then she pressed her dry, warm lips on mine. I felt her hand on the back of my neck, holding my head in position.

The train arrived with a hot roaring clatter. I opened my eyes and was amazed to see it was full of people who stood there watching us when the doors flew open. The words she'd whispered were "kissing cousins." Her lips had opened, and her mouth had tasted like fresh milk.

We never did get on the subway. We stayed on the bench while it clattered down into the tunnel toward Brooklyn, and she handed me my present. My hands shaking so I could hardly hold it, it gradually emerged from its wrappings: an English leather-bound appointment and address book from Dunhill's, with my initials tooled into the front—a pricey item indeed for a schoolgirl to buy with her own money. I remember staring at it for a long time while the smell of her hair surrounded me and I still tasted her mouth.

"Don't you like it?"

"I do," I said, turning it over and over in my hands and swallowing hard. "Of course I do."

"Well." She nudged me gently and kept her elbow pressed against mine. "Look inside."

There were alphabetical tabs for addresses followed by a memo pad for each day of the year. Automatically, I turned to the first letter of her last name. The entry, in her clear, round, prep-school handwriting, gave her name, my aunt's address, and underneath in block letters, "DON'T FORGET TO UPDATE WHEN NECESSARY."

"I bet you do," she whispered, leaning closer. "Forget."

"I bet I don't," I said. "And anyway, you're going to be there for a while."

"Who knows?" she said. "Do you think anybody knows what's going to happen to them?"

All I could do was stare in amazement at the sign on the other side of the subway tracks: an ad for Alcoholics Anonymous: TODAY IS THE FIRST DAY OF THE REST OF YOUR LIFE.

"Now look under today," she said.

I slowly turned the pages, not wanting the moment to end and apprehensive about what I would find. In the same block letters, it was simply: "CONGRATULATIONS FROM A SECRET ADMIRER."

I couldn't look at her. "Thanks," I heard myself say. "It's a really great present." I cleared my throat. "I wish . . ."

"What? What do you wish?"

She was smiling when I turned to her, the lock of hair over her eye. I really couldn't speak now at all. Did we kiss again? I've tried many times to remember, but I never can. Finally, I made my choice (made easier since I could now see from a nearby map that the D train bypassed this stop): "Listen. I think we should get a cab. . . . The next train might not come for hours."

When I stiffly rose from the subway bench and turned to look at her, my cousin's smile faltered in confusion. I let her go first through the turnstile, and opened the cab's door for her after I'd flagged it down. We hardly said a word all the way to Rockefeller Center. Her virtue stayed intact, and I do think that Kerouac might have approved.

In the fall, I went back to college as an English major. I lived in Cambridge, Massachusetts, where I fell in with a racy crowd, many of whom ended up at Warhol's Factory, including Edie Sedgwick.

PART 2

Paradise Island

Totems Revisited (1)

June 2004

NO PLACE IN THE WORLD has air like the Bahamas. I realize this unpacking my bags in my room at the Chaplin House bed-and-breakfast after thirty years away. The air is heavy with humidity, but not the oppressive kind I'd encountered as a journalist in Central American fever ports or in Southeast Asia. You can feel this air moving against your skin, salty, tangy, lively, cool. Your skin drinks it in and begins to feel supple and young.

I take this air into my hydrocarboned, post–9/11 New York City lungs and wander outside barefoot in shorts and a tee shirt the way I used to as a kid. The round cement stepping-stones set in the sandy brown soil are still there and feel exactly the same against the soles of my feet, a special warm roughness that I've never felt anywhere else. The handprint I made in one of them at the age of six is still there, too, impossibly tiny. I squat down and cover it with my own hand, which now looks like my father's: long-fingered, bony, and weathered from too much sun.

It's early afternoon. My three scientific colleagues—Dominique, Heidi, and Danielle—are busy unpacking their own bags in another cabin—in fact, the same one my sister and I shared in the old days. Ronnie and Joan Carroll have disappeared. Like a six-year-old, I roam aimlessly around the grounds, feeling the sun warm my face, scuffing my feet in the soft, powdery dust, picking a ripe sapodilla and sinking my teeth into the juicy flesh as a black-and-white-striped lizard on a higher branch pumps itself up and down on its forelegs and extends its red throat.

I have no doubt at all that the lizard is descended from the black-and-white-striped lizards I used to catch and keep in my room as pets. It makes me think of the little underwater ledge just off the beach where my father had first showed me my totem.

Fairy basslets can live for ten years. If there's one in residence now, it would be a great-great-great-grandchild of the original.

I go back to my room, change into swim trunks, and sneak out to the beach with my snorkeling gear, avoiding the house and concealing myself as much as possible in the shrubbery. The moment requires that I be alone.

The neighboring houses along the beach now sport high iron fences, NO TRESPASSING signs, and spotlights, but few tourists wander this far away from the Club Med or Atlantis, and the beach itself looks pretty much as I remember it. The water here on the ocean side of the island is almost as clear as it used to be: aquamarine to the reef a hundred yards out, then deepening to sapphire.

I put on my mask and snorkel and some free-diving weights, and wade out carrying my swim fins. Chest-deep, I bend my knees slowly as I did so many years ago, until my face is below the surface and I've passed through the looking glass into the silent world. A couple of silvery, plate-shaped gafftopsail pompano with their long, black-tipped dorsal and anal fins and forked tails circle curiously around while I pull on my fins. They feel like long-lost brothers.

Out in deeper water I take a few breaths, jackknife down, and ease along close to the rocky bottom, keeping my fins together and moving my body like a dolphin's. The rocks look exactly the same as I remember them, so do the little bubbles that form on the seaweed fronds, so does the orange pelican's tongue shell on the purple gorgonians. The prodigal son is home at last!

Cool, silky water eddies along the length of my body. Now I'm over a meadow of turtle grass with a small group of cuttlefish hanging in the water column ahead of me, waiting to nab an unsuspecting minnow. They dart away, changing from light, transparent blue to dark brown.

I'm right on course for the little ledge, even after all this time, and soon enough I can see it in the distance, a couple of silvery bar jacks circling over it. It's not too late: I can still veer off and avoid disappointment, but . . .

Here's my census for the ten-foot radius around the ledge: three male bluehead wrasse, one juvenile dusky damselfish, four bar jacks, one squirrelfish, one juvenile Spanish hogfish, one creole wrasse, one green razorfish, one blackbar soldierfish, four juvenile queen conch. There is no fairy basslet.

On May 9, 1959, forty five years earlier, my father took his own census on a similarly sized artificial reef he'd made from cinder blocks not far away (this had disappeared by 2004): one big-eye, many juvenile slippery dick wrasse, six juvenile brown chromis, several adult ocean surgeonfish, eight juvenile doctorfish, one banded butterflyfish, one red hind, one saddled blenny, one juvenile yellowtail damselfish, one margate fish, many young and several adult dwarf goatfish, one blackbar soldierfish, several mutton snappers, and many spiny lobsters.

There was no fairy basslet here either even in those days, but the mutton snappers, the margate fish, and the lobsters were important food sources not present in my count. My census turns up nine species, my father's thirteen. Now we come to the most interesting point of comparison between the two surveys: when he made his, the artificial reef was only three months old!

I'm rising slowly to the surface after my last look under the ledge when something big shoots over me with an underwater sound like ripping silk. I duck, wait as long as my breath will let me, then cautiously raise my head.

What I see is a black man in a yellow sport flotation vest astride a blue jet ski circling back toward me, waving one arm, mouth open in a shout. He doesn't slow down as he passes. What's he shouting? Could it possibly be "Watch out!"?

Ten years earlier, my closest friend was killed by a hit-and-run jet skier after surfacing from a dive in Shinnecock Inlet, Long Island. I raise my own arm out of the water to wave this man back.

"What did you say?"

My hand has become a fist. The jet skier turns away from me, the ridiculous waterspout of his machine's exhaust making him look like a scared white-tailed deer.

"*You* watch out!" I'm screaming at the top of my lungs. "*You* watch out, goddamn you."

Totems Revisited (2)

June 2004

"IT IS THE PART OF wisdom never to revisit a wilderness," wrote Aldo Leopold, the father of American wildlife conservation, "for the more golden the lily, the more certain that someone has gilded it. To return not only spoils a trip, but tarnishes a memory. It is only in the mind that shining adventure remains forever bright."

Of course, the central concept of this project flies directly in the face of that advice. I *expect* my memory to be tarnished, and measuring exactly the degrees and details of tarnishing is what my colleagues and I will be trying to do as we return fifty years later to my father's original collecting sites on this and future trips.

Leopold himself never heeded his own advice either. Conservation is based on keeping track of wilderness, taking a stand, not turning away. So I'm off to a pretty dramatic start: I'm almost killed trying to revisit my wilderness totem.

The most beautiful shallow reefs of my childhood were to be found off Lyford Cay, at the western end of New Providence Island about ten miles from Nassau. So, bright and early the morning after our arrival, taking Leopold's bull firmly by the horns, Dominique, Danielle, Heidi, and I pile into our rented outboard runabout to see what's happened to these reefs in the half-century since I last saw them.

The sun is behind us as I pilot the runabout west out of Nassau Harbor, which makes each coral head and ledge stand out sharply in the shallow water inside the reef. We're going at a good clip, maybe twenty-five knots, and the daily east wind has yet to pick up, so the water surface is glassy. We roar over the sand flats where Stan Waterman speared a huge tiger shark long ago, our wake leaving a reflected vee on the bottom. There's no sign of life down there now. Balmoral Island, an exclusive beach club in the old days, has been converted for tourist parties; its new name is Discovery.

Arriving at the Lyford Cay Marina channel, I hold close to the beach inside the reefs, and we drop anchor about where I had my second terrifying childhood encounter with a barracuda. Onshore, we can see a woman in a blue-and-white Lilly Pulitzer bathing suit saunter across the spacious lawn of her estate. Lyford Cay itself is now an exclusive gated community for tax exiles like Campbell Soup heir John Dorrance, investor Elliot Templeton, and actor Sean Connery.

Dominique, Heidi, and Danielle are still fiddling with their SCUBA equipment, cameras, clipboards, and measuring gear as I go overboard backwards in a cloud of bubbles. When I get my bearings and can look around, it takes a few moments to understand exactly what I am seeing. Finally, it comes to me: the light has gone out.

It is a sunny day, and plenty of light shines through the surface onto the reef, but it is absorbed like light on a winter forest. Dark green-brown algae covers the broken branches of elkhorn coral, and they no longer glow with that magnified, intensified terra-cotta fluorescence. Under the algae, the coral has died.

You can read about this destruction, and I have, but that doesn't even come close to preparing you for seeing it firsthand. I swim around the "bare ruined choirs" in a daze, trying to remember why I'd come.

Science! The three scientists are now in the water with me, consumed in their work. Dominique, holding a clipboard with a waterproof form attached that lists all fish species likely to be seen, is recording population estimates ranging from "rare" (one specimen) to "abundant" (over one hundred) for each species. She'll be counting for fifteen minutes throughout a circle thirty meters in diameter, beginning in open water above the reef, with free-swimming fishes like snappers, grunts, and chromis, then free-diving down (tank-diving when pregnant is a no-no) to check the crannies in the dead coral that shelters cardinalfish, gobies, and blennies. Heidi, using SCUBA, is laying out a thirty-meter tape over the top of the reef, down its side, and along the bottom nearby. Later she'll video and photograph along this transect. Back in the lab she'll analyze the images for type and percentage of cover: rock, sand, algae, live coral, et cetera. Danielle is sampling the water with a sterile polyethylene container to test later for suspended particles that can show the level of pollution.

Gathering data is comforting, but still . . . they didn't see these reefs fifty years ago. I grab a clipboard and form and do a fish survey of my own. Dominique and I between us count very few grunts, no tiger groupers (or any kind of grouper), no snappers. Neither do we count any spiny lobsters, eagle rays, drums, filefish, toadfish, soapfish, or cherubfish.

It was revisiting wildernesses, witnessing firsthand how they'd changed over time, that set Leopold on his life's course. But I don't think even Leopold saw such a drastic example of deterioration as I do at Lyford Cay on this first day of our explorations: 90 percent of his beloved Southwestern forests did not die during his lifetime.

In his seminal essay, "Thinking Like a Mountain," from *A Sand County Almanac,* Leopold writes that his epiphany came while watching a wolf he'd shot:

> We reached the old wolf in time to watch a fierce green fire dying in her eyes. I realized then, and have known ever since, that there was something new to me in those

> eyes—something known only to her and to the mountain. I was young then, and full of trigger-itch; I thought that because fewer wolves meant more deer, that no wolves would mean a hunters' paradise. But after seeing the green fire die, I sensed that neither the wolf nor the mountain agreed with such a view.

The barracuda is the wolf of the reef, but my own epiphany is more zen. I don't shoot a barracuda and watch it die, I just never see any barracuda at all (at least not over six inches long). Not then, nor in the following ten days of diving up and down the north coast of New Providence Island as we scout out and survey as many of the old collection sites as we can find. This is a recon trip, to establish the lay of the land, to suggest the overall direction of the project. The actual collecting of fishes will be done on subsequent ones. If we can get the rotenone permit.

Underwater, off Lyford Cay, I keep looking into the blue distance for that spine-tingling, silvery glint, and when it doesn't appear I realize how much I miss it. "Once seen, never forgotten," as my father lovingly described the big, scary predator in his *Fishwatchers Guide*. Certainly true for me. My first sighting in the little cove at Treasure Island is seared into my brain forever. He respectfully titled any barracuda over five feet an "old growler," and relished telling the tale of a southern Bahamas fisherman who encountered one that was longer than his twelve-foot dinghy. This was the only fish he immortalized, in the mosaic and the oil painting that now reside in my New York loft. He adored the frisson of barracudas. Pretty strange to imagine that the creature that had terrified me most as a child could easily be his totem.

Hope Springs Eternal (1)

June 2004

Even at age seventy-one, Sir Nicholas Nuttall's face lights up exactly as my father's would have on sighting the three lady scientists in their swimsuits. I know the look well: it's one of appreciation, not lasciviousness, though there is a twinkle somewhere. It's as if my father has come back to life, Old Etonian accent, family regiment, sporting proclivities and all—though, in fact, Sir Nick is not that much older than I am. He's had several wives, one of them an ex of Peter Sellers.

Like me, Sir Nick is trying to build on my father's work in the Bahamas. They would have loved each other. I can't help feeling a twinge of jealousy. Here I am desperately trying to redeem myself and live up to a legacy that Sir Nick already inhabits. It seems to fit him perfectly, and he seems much more qualified than I—a diver, mountaineer, and marathoner with a distinguished record as a soldier and businessman, who made his family company hugely profitable before selling out to become a tax exile. Oh well. He accepts with a little bow the copy of *Fishes of the Bahamas* that I have brought along for him.

"Thank you so much. This is the bible here, you know."

My father was a founder of the Bahamas National Trust, set up in 1959 to administer the world's first marine protected area in the Exuma Islands southeast of Nassau and to oversee new conservation projects. The Trust had the blessing of the white colonial "Bay Street Boys'" administration, and things went very smoothly for the effort.

In 1993, Sir Nick founded the Bahamas Reef Environment Educational Foundation (BREEF) because "the Trust had rather been taken over by commercial interests, you know." For an English expat in a newly independent colony, things have not been nearly so easy as they were for my father. He tends to be seen by Bahamian authorities as a bit of a gadfly.

Still, BREEF has successfully campaigned for a closed season on the declining Nassau grouper, for one, and even this early on, Sir Nick claims, you can see more young grouper on the reefs. For another, he says, *Diadema* are coming back after a near total wipeout from disease. *Diadema*, the poisonous black spiny sea urchin, are crucial to reef health because they eat coral-destroying algae. For a third positive sign, he has noticed that new elkhorn coral is beginning to "recruit" on the outer reefs around Goulding Cay near the edge of the drop-off into the Tongue of the Ocean. He's invited us aboard his skiff on a voyage of hope.

Like most of his fellow tax exiles, Sir Nick lives in an exclusive gated community near Lyford Cay that happens to have been built over a drained mangrove swamp—one of my father's most productive collecting areas fifty years ago. We would seine the creek with a fifteen-foot net, catching juvenile fish of all kinds, since the mangroves are an important fish nursery. Now a system of sterile canals connects the villas to the nearby ocean. If Sir Nick sees the irony here, he doesn't mention it as we clamber aboard. His smile has more to do with the scenery.

My three colleagues, in spite of (or perhaps because of) their impressive credentials, do present an arresting spectacle. Dominique's maternity maillot—she is now five and a half months pregnant with her second child—is yellow with electric blue polka dots, Danielle's two-piece reveals the lean, wiry body of the surfer she was in another incarnation, and Heidi in a navy tank suit looks more like a rugby champion than ever.

Also with us are BREEF's comely director, Casuarina McKinney, twenty-seven, and her equally comely younger sister, Taja, great nieces of my father's good friend and Bahamas National Trust cofounder Andy McKinney. In the skiff, heading out to the ocean, the five women discuss the latest coral reef studies (Casuarina studied marine biology and environmental policy at Duke) and how the Chaplin Project, as we are bold to call it, will fit in. I stand beside Sir Nick at the console, and he looks over with a smile. "I find women are especially eager to learn, you know."

Just so. On later expeditions, we will find that women students make up the bulk of the interns we use in the field. All the better, as my father would have said.

As we roar across the shallow bay between Sir Nick's house and Lyford Cay itself, I notice the water has become vegetable green, a big change from when we dove here a few days earlier. "Sewage seep," says Sir Nick concisely, and explains that most of New Providence Island's ballooning population still uses septic tanks that overflow in rainstorms, like the small one yesterday. Also, the Nassau sewage system pumps its waste deep into the limestone bedrock through leaky pipes. Water samples taken here by Danielle, in fact, will show a significantly higher proportion of the large suspended particles that indicate pollution than those taken farther east (upwind).

At the very end of Lyford Cay, what looks like an enormous Mayan temple is under construction. It's the home of Canadian women's clothing magnate Peter Nygård. "They never finish." Sir Nick shakes his head in wonder. "Truly fascinating. Something new every time I go by. Look there, will you? Is that an elevator? Or is it a swimming pool?"

As we cross the bank to Goulding Cay, the water gradually changes from the soupy green to something approaching undiluted Pernod. We drop anchor in a patch of white sand surrounded by the dark brown of dead coral. Because of the island's proximity to the Tongue of the Ocean, the Bahamas's answer to the Grand Canyon, the shallow reefs here in the old days contained more fish than any others around Nassau. The barracudas were bigger; Stan Waterman said he'd seen tiger sharks. It was the wildest place available on short notice.

But today I'm the only one who has lived this history. I wonder what the others are thinking as we snorkel after Sir Nick and Casuarina through tragic stands of crumbling elkhorn. Maybe, to them, things could be worse. To me, hope seems very far away. That's the trouble with having a fifty-year baseline.

I'm checking apprehensively for the loom of those big barracudas and tiger sharks from fifty years ago, but all I see is Casuarina, in a well-cut floral print two-piece bathing suit, swimming ahead of me. As I watch her, I can hear my father's voice as clearly as if he were speaking into my ear: "A toothsome morsel, what?"

The largest fish we encounter today are foot-and-a-half-long yellowtail snappers circling the area where dive boats feed them up for tourists. We see no groupers of any kind, in spite of the closed season. And the branches of new elkhorn that Sir Nick points out so proudly seem terribly incidental when compared to the total destruction all around them. Finally, we can find only one of the six lowly but crucial *Diadema* he said he saw here earlier. Sir Nick's message of hope falls flat as a flounder at least as far as I'm concerned.

My father would have taken exactly the same stiff-upper-lip, English tack as Sir Nick does: westerly gales might strip away the sand on the beach in front of the Chaplin House, but don't worry, the prevailing easterlies will bring it all back in no time. They both went to the same school and speak with the same accent. Conservation, of course, used to be all about preserving the status quo.

I don't think it can be anymore. My favorite outdoor writers are Edward Abbey, Jim Harrison, and Carl Hiaasen, whose idea of conservation is to blow up dams, free the rivers, and feed tourists to alligators. The status quo has turned into a grave threat.

My fledgling activism is put to the test a few days later. While the scientists meet with various authorities in town, Ronnie drops me off to rubberneck around the huge Atlantis resort at the other end of the island.

But admission to the three-million-gallon outdoor aquarium (world's largest) is $35 for non-guests, and I balk at paying it. Every single slot machine in the Caribbean's largest casino is occupied, and the quarter-mile Lazy River tube ride is full of screaming kids. So after a slow hamburger at the Lagoon Bar & Grill, I decide to walk back to the Chaplin House along the beach.

Atlantis is in the process of expanding in the same direction. At one of the vast construction sites, a chain-link fence reaches almost to the water. Circling around it, I'm stopped by a security guard: private property.

"But I live up there," I say. "I'm just trying to get *home*."

Something in my voice prompts her to radio for a backup. Good. I hope I'll get arrested. Arrested trying to go home along a public beach.

When the backup unit arrives in a golf cart, I announce I'm going home this way no matter what they say. They just stand there and watch me walk away.

Beneath Atlantis

June 2004

ATLANTIS FASCINATES ME, BUT NOT for reasons foreseen by Sol Kerzner, its South African developer. On my rubbernecking expedition, I found that the main complex is built on pilings over the same canals that were among my father's most productive collection sites in the old days. I want to dive there.

The canals were originally dredged in the thirties by Swedish industrialist Axel Wenner-Gren, who'd bought the whole of eastern Hog Island from the financier Edmund Lynch for $150,000 and developed it as his personal Shangri-la. Wenner-Gren was dogged by rumors of being a Nazi sympathizer, and for some time locals suspected the canals to be a refuge for German subs (they weren't deep enough).

Our neighbor Huntington Hartford bought the property from Wenner-Gren in 1961 for $9.5 million, changed the island's name from Hog Island to Paradise Island, and planned a luxurious resort-casino that never got off the ground.

Other owners, including Donald Trump and Merv Griffin, came and went. Finally, in 1994, South African developer Sol Kerzner put in the present-day Atlantis at a cost of some $200 million.

No doubt my father would have loved the Atlantis aquarium. There's live coral, more than a hundred species of local fish, and the Predator Lagoon where you can see more than one hundred sharks. Behind the thick glass walls, things look the way they used to in his day, only better. Things could even look close to the way they were before mankind arrived on the scene. You can take it all in from a table in the Café at the Great Hall of Waters, while eating one of the declining grouper population.

On a day too windy to work comfortably in open water, I pilot our runabout into the canal leading to Atlantis's sixty-three-slip marina, filled with luxury yachts up to 220 feet long. The twin bronze marlins on the crest of the Royal Towers (Nassau's tallest building) soar above us as I drop anchor out of the main channel in front of the heavy iron portcullis that shields its underpinnings from public view. Dominique and Danielle are kindly indulging my obsession to get below the surface, but Heidi has no interest in visiting Atlantis even to dive under it. She's back at the Chaplin House working on her notes.

A black Bahamian idles past in his own skiff, smiling and waving. "He probably wants to know how you happen to be alone in a boat with two beautiful women." Today, Dominique is resplendently pregnant in Scottish tartan. "He wants to know your secret, Gordon."

I spread my hands out palms up and shrug. "Dumb luck?"

"Naw," Danielle says. "We're *scientists*. Good solid planning is what we specialize in. Luck doesn't come into it at all."

It's not clear how deeply the portcullis extends down into the water. "Looks a little spooky for a girl in my condition," says Dominique doubtfully.

"I think we can make it." I'm already putting on my gear.

The surface of the canal is slick with oil and gasoline from boat traffic to the marina, farther past the huge pink building. Everything above water, in fact, is pink except for the black portcullis. "I'm not swimming in that," Danielle says. "It'll *ruin* my hair."

"Okay. I'll reconnoiter." I lower myself from the dive platform into the vegetable-green water and swim slowly toward the portcullis, waiting for a shout from the security people.

At first glance, the only living things seem to be pulsing sea anemones that look like brown saucers on the muddy marl ten feet below me. But wait! I can't believe my eyes. A little bit ahead is what looks like a living, full-grown queen conch, one of the Bahamas's national treasures. Conchs can live for twenty years and have traditionally been a prime food source in the islands. In 1992, the Convention on International Trade in Endangered Species listed the conch one category below "threatened."

This one must be dead. I dive down and turn it over. It's radiantly alive, the black foot with its two stalked eyes searching for purchase, the bright polished pink of its lip matching the pink of Atlantis itself. Literally the last thing I expect to see here.

I gently place the conch back where it was. A fluke, that's what it must be. But here's another. Maybe the larval conch veligers escaped from the aquarium inside through the water pumps? Wonders never cease.

By now I can see that the portcullis only extends about a foot and a half below the surface. On the other side is a huge dark cavern punctuated by thick concrete pilings. The *plink* of droplets from above echoes through the void.

Under the radar! I wave to the scientists and slip below the sharp points of the portcullis. It takes a little while for my eyes to get used to the gloom inside, but at the edge it's not too much different from the gloom of thick mangrove roots and the weed that attaches to them.

I begin to pick out the tiny fish one at a time. A six-inch-long barracuda. Little schoolmaster snappers, with that severe frowning bar over the eye. An orange-eyed checkered puffer or two. Yellow-and-blue-striped grunts, hogfish, baby groupers, juvenile rainbow, and indigo parrotfish. Suddenly I realize that they're all here, all "our old friends," as my father used to say—all the young fish that typically mature in a mangrove environment. To them the pilings of Atlantis are nothing more than a substitute for mangrove roots, and apparently not a bad one.

Death of a Scientist

June 2004

IN THE EVENINGS, AFTER A day in the field, Dominique, Danielle, and Heidi like to sit out on the verandah of their cabin, review their data—fish surveys, habitat assessments, water samples—and write up their field notes just as my father and Jim Böhlke had done fifty years earlier.

They've brought with them, in fact, Xeroxes of all those notes from the old days. While the scientists work, I read through them in my room, and their energy and painstaking commitment never fail to amaze me. It's a side of my father I never experienced before, though I was vaguely aware that it existed. And though I knew the end result.

The writing is usually clear and concise, but sometimes my father gets carried away and his real voice comes through:

> In the stomach of the wahoo, which was otherwise empty, were two revolting living parasites. About 1 inch long, the same colour and general appearance as a newly hatched

My father, with my sister Susie in his lap, and me—with mounted moths!—on the seawall of the Chaplin House, Nassau, 1951

Mr. and Mrs. Charles C. G. Chaplin, August 19, 1937

South-side verandah of the Chaplin House, 1951

North-side beach, 1951

My mother, Louise Chaplin, on the beach at the Chaplin House, Nassau, 1951, and in typical beach attire, 1958

Still life with palm frond, 1952

Boy snorkelers, 1953

My father on the prowl, 1953

Hogfish dinner, 1953

Jim Böhlke, my father's mentor, on the Amazon, circa 1960

My partner Susan, black sand, white water, Hawaii, 1992

Diseased coral, New Providence Island, 2005

Loren Kellogg mixing fateful rotenone, 2006

Dr. Katriina Ilves with a porcupinefish specimen, 2010

Katriina bags a porcupinefish, 2010

Sorting table with Dr. Ron Eytan at work, three a.m., 2010

Katriina's crew, 2010

Susie's triumphant arrival by paddleboard and our reunion, Saba, 2009

> sparrow. They had long prehensile necks which could be extended another inch and which constantly weaved about in a blind but sinister manner. At the end of this neck was a mouthlike orifice. Underneath the neck, on the main body, was another orifice of unknown function. I placed them in a beaker of salt water where they seemed fairly happy, exuding drops of what looked like digested blood.

Who would compare a revolting parasite to a newly hatched sparrow? Or take such apparent pleasure in the blind, sinister manner of their neck-weaving? Or note that they seemed "fairly happy" exuding those drops of digested blood? Only a self-taught Englishman with a quirky sense of humor who loved to read his young son ghost stories.

I join the three on their verandah after the chores are done. Scientists love data the way writers love words, the building blocks of their professions. But I've noticed that for many scientists, unlike for writers, gathering data is an end in itself. What it is used for later, if anything, is completely beside the point. As far as it goes, I too find the scientific process calming, reassuring, and familiar—the careful entering of the data, the stories about the day's work, the jokes. Earlier that day Dominique had snagged her ample bathing suit on a deck cleat while trying to slide overboard and had hung there helplessly for a while until the material ripped. Everyone else was underwater. "I called and called, and nobody heard."

"Oh, we heard all right," Heidi says. "But you wouldn't have wanted us to interrupt our work, would you? Plus we knew the suit wouldn't hold forever."

"Well, now the suit's finished. Is that what you wanted?"

"It's not finished," Danielle says. "Just delightfully risqué. Right, Gordon?"

I shrug and grin.

"Look at him blush. He has a fetish for pregnant scientists."

Later, I try to get a better feel for the direction our work is taking. It all seems very random at the moment. Exactly how are all our painstaking fish counts and bottom surveys going to be used? The list of collection sites is formidably long. . . . We can never pinpoint all of them. How are we going to cut it down to a manageable size? "New trends will come out once all the data is analyzed," Dominique says soothingly. To my nonprofessional ears, it sounds a tiny bit like mumbo jumbo.

Danielle, who put together the comprehensive spreadsheet we are using of hundreds of Böhlke-Chaplin collection sites, has her own take on our project's lowly status. "I'll tell you one thing: if three other people at the academy had proposed this, the results would have been different." She is talking in part about her sex change and how it has affected her career.

"For eight years, as a man, I was the biggest fund-raiser in my department. In the last two, I've been cut out of the loop. I'm only part-time now, and I'm underfunded."

Her personal life does sound complicated (or uncomplicated, depending on how you look at it): she's still with her original spouse and their seven-year-old son, conceived and born before the change.

We open a bottle of wine and start making preparations for dinner.

"I'm not one of those pregnant ladies who doesn't touch a drop," says Dominique, although she vows she never drank alcohol at all before she got her doctorate and became a full-fledged ichthyologist.

"Ichthyologists tend to drink a lot and then shoot themselves." Heidi cuts her eyes at me. "Right?" She knows of at least two such cases at the academy.

I know about one of them: my father's coauthor, colleague, and friend Jim Böhlke.

I was about ten years old when Jim first showed up at the Chaplin House, and I have to say, I wasn't very impressed: fish-belly white, overweight, earnest, hardworking, a total nerd. He lived in unfashionable baggy Hawaiian shirts and khaki shorts. Like many scientists, though, he had a mordant sense of humor, and his enthusiasms were boyish and infectious. Plus he knew more about fish than anyone I'd ever met. My father's deference was obvious and intriguing.

Oddly enough, Jim's interest in fish, like my father's and mine, had begun with an image. His was a mottled brownish South American armored catfish that was the only fish left alive in his childhood aquarium after the aerating system gave out and the others were asphyxiated. The armored catfish is a bottom-feeder, using its long, sensitive barbels to stir up edible detritus that otherwise would pollute the environment, but it can also breathe air from the surface. At the tender age of fourteen, Jim wrote to Henry Fowler, curator of fish at the academy, for more information about this miraculously hardy fish, and Henry Fowler *answered.*

That was the start of everything. After a stint at Stanford, studying with the famous George S. Myers, Jim found himself at the academy working in the ichthyology department under Fowler and with my father, who was now officially a research associate. It was a long way from Buffalo, Minnesota, where Jim's father ran a hardware store. I imagined his family as the kind of sober German Lutherans who populate Lake Wobegon.

It never occurred to me that Jim was young enough to be my father's son and my older brother. Dissimilar as they were in age and background, they spoke the same language—with different accents—and both had the same look on their faces when they surfaced from a dive, even after eight hours underwater: wonderment. But there was never any doubt that Jim was the senior partner.

Jim's drive and ambition came across to me most strongly through his hands. To preserve the specimens we collected, he and my father worked with a highly toxic preservative called formalin. Jim thought protective rubber gloves, routinely used now, were too clumsy: he always sorted specimens bare-handed. By the end of a collecting expedition, his hands would be fissured with deep feathery cracks like glacial crevasses, red instead of blue. I couldn't take my eyes off them, but he himself ignored them.

How could he do it? In his monomaniacal pursuit of more specimens, in fact, he seemed totally unaware of all physical hardship or danger. I can see him now at Grassy Creek, Andros, submerged with primitive SCUBA gear, in khaki pants and a long-sleeved Oxford button-down shirt to protect his blistered, peeling skin. His attention is completely concentrated on a fish the size of his little finger, and when he returns to the larger world it is to look calmly from one eye to the other of that hammerhead shark.

Over the next twenty years of hand-in-hand work, some of it in the field but most of it in the gloomy ichthyology lab in the basement of the academy, the Böhlke-Chaplin partnership turned out to be inspired. "Thanks to Jim, Charlie actually *became* a scientist," according to his friend and colleague, the famous and daring Dr. John E. (Jack) Randall of the Bishop Museum in Honolulu. I knew very well how much my father valued this alchemy but didn't at all understand how it had happened. After all, he'd never had any formal training; he'd never even been to college.

How hard he worked at it was emphasized for me in a review of *Fishes of the Bahamas* when the finished work appeared in 1968: "Despite the fundamental importance of accurate identification as a basis for all biological work, the production of good faunal studies is singularly unrewarding. If the job is well done, it looks so simple that the years of collecting, literature searching, laboratory comparisons, testing of keys, and writing and rewriting of the text can hardly be appreciated by anyone who had not attempted such a work."

This painstaking and inglorious process elevated Jim himself to one of the top authorities in the tiny world of ichthyology, with more than 120 publications to his credit as well as the presidency of the American Society of Ichthyologists and Herpetologists.

As a kid, I was pretty much unaware of Jim's paralyzing shyness. I did notice he blushed and was quiet around company, but that's the way

I myself behaved. Sometimes we'd catch each other's eye and grin: here was an ally. I had no idea that before he gave a speech, presented a paper, or hosted a party he had to fortify himself with alcohol.

To protest his eventual firing, Jim's friend and supporter, the eminent C. Richard Robins, then director of Miami University's Division of Biology and Living Resources, wrote academy officials: "The Academy has always found it necessary to entertain visiting dignitaries and potential donors, and the Directors have spread out this task among the various curators," his letter pointed out. "In particular, Jim has done much entertaining on behalf of the Academy. On at least two occasions during my visits he was asked to entertain visitors and members of the Board of Trustees at his house. On several occasions, and again during my visits, large-scale cocktail parties were held within the Academy. It should not be surprising that under such circumstances someone becomes alcoholic." Robins himself had turned down the directorship of the academy because of these responsibilities.

There was very possibly a tragically ironic dimension to Jim's problem, stemming from those fissured hands in the old days. Much later, when I mentioned to his widow Genie the possibility of formalin toxicity weakening his defenses and making him more vulnerable than average to alcohol poisoning, she agreed that it existed but didn't want me to write about it. She's dead now too. So here it is: I believe there is a good chance that Jim literally sacrificed himself for science.

According to the Labor Department's Occupational Safety and Health Administration (OSHA), ingestion of as little as thirty milliliters of formalin can cause death, not only by damaging the stomach but also the liver, kidneys, spleen, pancreas, and brain and central nervous system. Long-term exposure can cause cancer of the nose, sinuses, pharynx, and/or lungs, as well as acute bronchitis, asthma, nasal tumors, hives, eczematous dermatitis, and epithelial cell changes. OSHA standards for working with formalin mandate not only rubber gloves but protective clothing and respirators when used in enclosed spaces.

Jim spent far more time working with specimens in the lab than he did in the field. He may have worn gloves there, but never protective clothing or a respirator. The other deceased academy ichthyologist that Heidi had mentioned no doubt had had the same work habits. They all do, even Dominique.

The academy sacked Jim in February 1982, not long after his fifty-second birthday, citing "a number of years of continued absences, only occasional bursts of productivity, and a current lack of adequate contributions to the academy research program." According to a staff memo, he had been turning up for work with alarming cuts and bruises attributed to "car accidents and accidents at home," and had once needed the assistance of a security guard to make it through the front door of the building and into his office.

When he heard the sacking had actually taken place, my father, seventy-six at the time, wrote worriedly to the academy president, Dr. Thomas Peter Bennett: "I feel it would be the end of Jim if he were not given the chance to prove himself."

To cover himself, Bennett detailed in a subsequent staff memo how my father already knew that the action was being taken only after two years of administrative concern, ultimatums, and academy-financed rehab programs.

But my father was also aware (unacknowledged by Bennett) that Richard Robins had recently enrolled Jim in a new and intensive rehab program "with great success," and that Jim had finally faced up to his problem after years of denial. In his impassioned plea for reinstatement, Robins wrote the academy that "in view of these happy developments it was astounding and very discouraging for us to learn of the Academy's action to dismiss him. I cannot understand why such action would be taken at this time and under such circumstances. . . . Jim is worth fighting for!"

The academy stood firm. Jim's twenty-eight-year career there was finished. On March 25 his wife Genie found him dead in his home laboratory with a .38-caliber pistol bullet through his temple.

"Jim was supposed to come and have a talk with me on Tuesday," my father wrote their mutual friend and colleague Jack Randall, "but never showed up and I shall always wonder whether I could have cheered him up enough to tilt the balance in favor of carrying on and realizing his value and potential as a top-notch scientist. Of course the Academy might have handled the problem differently."

Could my father have thrown his weight around more effectively to prevent Jim's firing or to have him reinstated? Well, yes, he probably could have. After all, my mother had endowed the Chaplin Chair, which Jim and all subsequent curators of ichthyology at the academy have held.

But throwing his weight around had never been my father's style. How did he feel, having predicted Jim's death, when it *actually happened*? His letter to Randall said as much as he ever allowed himself to say about such things.

Like my father, Jim requested in his will that his ashes be spread on Bahamian waters. His family chartered a small plane in Florida, flew across the Gulf Stream to the Great Bahama Bank, and let the ashes loose through an open cockpit door.

Ashes

June 2004

I'VE HAD MY FATHER'S ASHES with me for thirteen years, in a pink plastic box about eight inches square, designed to look like marble. I know he would have hated the box, but I haven't replaced it because it is only temporary. In his will, he requested the ashes be spread at sea. I've always known exactly where.

I want to be alone when I return to this place for the first time. On the other hand, it was my father's most treasured collecting station. I weigh the importance of taking the scientists there and including it in our survey against the fact that I need to consecrate it as my father's final resting place. I decide that the latter takes precedence. I can wait until after they leave, since I have an extra day.

By the end of our trip, the trio of lady scientists seem happy enough with what we've accomplished. We've rediscovered at least some of the old sites, have taken water samples on them, made visual fish counts and bottom surveys. We've mapped out the logistics of getting to these sites. We've established contacts in Nassau like Sir

Nick and a few government officials (though we have yet to meet with the Fisheries director to ask about the rotenone permit).

Our preliminary observations show that coral destruction worsens downwind and down-current (east to west) from the city of Nassau, and is at its worst at the western end of New Providence: Lyford Cay, Goulding Cay, et cetera. Danielle's water samples in these areas show definite pollution. And we've unearthed a video transect taken in 1997 at Goulding Cay showing good coral cover just a few months before a drastic El Niño condition raised the water temperature more than one degree centigrade.

It's a start. But the effort that will be required to shape these fairly random observations into a rigorous study and the time that might take give me a sinking feeling in spite of the scientists' good spirits. They have other commitments. The rotenone permit for future collections is still up in the air. And Dominique will soon have another baby to take care of.

Last but not least is my dawning awareness that I have far more riding on this project—redemption, paying back debts, carrying on my father's legacy to me, just to name a few—than the scientists do. Has this just been a lark for them? A trial balloon? My awe and respect for scientific inquiry, in particular by the Academy of Natural Sciences, inclines me strongly to give them the benefit of the doubt, but still. . . .

I wave them off from the Chaplin House dock not realizing that I'll never work with Dominique and Danielle again.

Right after they leave, I get a surprise phone call from Sir Nick, who has been trying to arrange a meeting for us with the Fisheries director, Michael Braynen. "Can you meet me outside the Fisheries Department in an hour?"

Braynen, slim, black, and dignified, wears a dark suit that helps complete his aspect of a wearily hip jazz saxophone player from the fifties; Sir Nick sports a beautifully cut English houndstooth tweed. This is, after all, a former British colony. Aides scurry in and out of the conference room, and I sit there in my khakis and sport shirt feeling terribly underdressed.

After I finish my little presentation, Braynen puts the tips of his fingers together and looks at the ceiling. "And what are the lasting effects of rotenone, can you tell me?"

"We never saw any. And we collected at the same places many times. We'll be happy to forward you the most current information on it. And, of course, after the collections we'll make all our data available to you."

He tilts his head in my direction. "I know your father's work, of course."

"He'd be glad to hear that." I brought a copy of *Fishes of the Bahamas* and now pass it over to him. "He would want you to have this, I'm sure."

"Oh." Braynen smiles. "Well, we already have a copy, of course."

I reach for the book to take it back, but he gently puts his hand on it. "Thank you, we'll be happy to have this one too. It's bound to come in handy."

The meeting is concluded with his cordial invitation to apply formally for the permit.

Back on the broiling sidewalk outside the department offices, Sir Nick raises his eyebrows and gives me a slightly rueful smile . . . as if we're two schoolboys after a disciplinary visit to the principal.

"That seemed to go quite well. Didn't it?" I ask.

"Yes . . . I thought so."

"Thank you so much for arranging this."

"Well, he's quite a reasonable chap, you know." Sir Nick is referring to the fact that Fisheries recently initiated the three-months' closed season for the Nassau grouper that his organization, BREEF, had campaigned for.

"Do I have reason to be cautiously optimistic?"

"I'm always optimistic. It's the only way to be." This statement comes one year after he ran the New York City Marathon to raise funds for BREEF and two years before his death from lung cancer. "But don't tell Casuarina."

Casuarina has made it clear that she's strenuously opposed to our use of rotenone.

The pink plastic box containing my father's ashes sits on the skiff's floorboards next to my feet, so I can stop it from sliding if I hit rough water. I know the route like the palm of my hand: out Nassau Harbor to the east, through the channel between Hog (now Paradise) and Athol Islands, where the coral sea gardens used to be, across the deep shark channel between Treasure Island (now Salt Cay) and the rocky little cays just west of Rose Island, and finally across the reef-studded bay north of Rose to where Green Cay rises scrubbily above gray limestone rocks and a white sandy point backed with palm trees and sea grape. The roof of an abandoned house shows through the green. It was there fifty years ago. Nothing seems to have changed. The sky is overcast with high cirrus, and a light easterly trade wind is blowing at about eight knots.

I head out to sea around the cay's western point and slow to a couple of knots two hundred yards offshore. Peering over the side, through the sapphire of moderately deep water, I check for the dark loom of coral heads against the lighter sand bottom. My father would never forgive me if I picked the wrong one: he loathed sloppy seamanship.

I hadn't been with him when he died in the spring of 1991. Ironically enough, my partner Susan Atkinson and I were voyaging across the Pacific in a small sailboat, among other things to fulfill his childhood dream of circumnavigation, cut short in Barbados when he married my mother. We heard the news from my stepmother over High Seas Radio at nightfall several hundred miles south of the tip of Baja California. "You must remember that he was very proud of you," Susan said. "Even if he didn't always say so."

Alone on watch that night, I waited in vain to hear my father's voice, to see some kind of sign from him. Susan had the dawn watch;

I was in my bunk dreaming of flying. Or maybe swimming. When I heard her scream from the cockpit, I rushed on deck expecting I don't know what.

Her eyes were huge, and she was waving her arm to windward. All the way to the horizon, dolphins were leaping from the water, hundreds of them. The spectacle seemed to last for hours. My father, who loved to confound my mother's friends by telling them he was a firm believer in reincarnation, had always maintained he'd return as a dolphin.

A few days later, motoring through a glassy swell near Clipperton Island, we encountered a small pod of pilot whales, four or five cows and a much larger bull with a straight-up dorsal fin like a killer. The bull left his cows to come after us, and I remembered that one of my father's favorite survival tales was about a family who spent months in a dinghy after their boat was sunk by pilot whales not too far from here.

The bull held ten feet astern until we finally went below for lunch. Then we felt a jar, as if the boat had run aground on a soft reef. We rushed to the cockpit to see the bull's round, black head less than a foot from our big rudder. If it were disabled, we'd have no steering hundreds of miles from land. We held our breaths as the bull closed the distance and butted the rudder gently. Then he sheered off and we never saw him again.

We thought this was the best of omens: my father's gentle send-off and blessing. Ironic, no? Considering how it all ended a few months later in the Marshall Islands, when Susan and the boat were lost in Super Typhoon Gay.

The widely separated coral towers are dotted around a fifty-foot-deep sandy flat, a few rising to within fifteen feet of the surface. I try to put the skiff into the right historical gestalt—triangulated from the Green Cay point and the ruined house at the correct distance from shore and the right depth—and search for a particularly high, narrow tower. For inspiration, I slide my bare foot over so it's touching the pink plastic box. My hand grips the wheel so tightly it cramps. I curse the high cirrus, which mirror off the surface, making it harder to see down. I wish I had someone to tow me through the water while I searched, but then I wouldn't be alone.

Finally, I drop anchor near a likely-looking head, pull on my snorkeling gear, and jackknife backwards over the side. Light slants down in shafts onto the intricately layered, castellated coral structure, but the fact that most of it is dead is secondary to the fact that it's not the one. It's not high enough, and it's not shaped right. I'm positive, even though it has been thirty years.

Up anchor, move boat, drop anchor, jackknife overboard. A few more unsuccessful tries, and I'm beginning to wonder whether my mental picture is as accurate as I'd thought. Or if maybe the head has decayed and crumbled. I stretch the gestalt to search farther afield. I start to think in terms of compromise: well, if it's not the exact same head, at least it's in the exact same *area*.

I'm floating tiredly on the surface, looking down at a tall, stately tower crowned with a big knob of half-alive brain coral that looks vaguely familiar (in the old days, the entire knob would have been alive). I try a shallow dive to the top, peering around while the blood lags in my brain, and recognize the declivity that my father disappeared into fifty years ago while I watched.

The consistency of ashes is not dissimilar to the consistency of powdered rotenone. I've brought a pillowcase along, and back in the boat I unscrew the plug in the pink plastic box and pour the ashes into it. "Well done, Gordy," a familiar voice whispers inside my head. "What on earth took you so long?"

My plan is to dive to the declivity and let the ashes free inside it. It begins about thirty feet down, a challenging dive for me these days.

I put an extra two-pound weight on my belt to give me slight negative buoyancy and jackknife over the side with my father in his pillowcase. One secret of free-diving is to relax and focus your mind on something else so that you arrive at your destination unexpectedly instead of struggling for it. Another is to breathe deeply and regularly on the surface, filling and emptying your lungs completely but not hyperventilating—which can cause the dreaded "shallow-water blackout" by fooling your body into thinking it has more oxygen than it actually does. A third is to clear your ears before they start to hurt: in my case, once on the surface, then every twelve feet. It took me years to discover that I have to turn my body upright, hold my nose, and blow—I can't clear head down. But the process is different for everybody: Stan Waterman could swim straight down at full speed swallowing as he went. Eustachian tubes like "Grand Prix exhaust pipes," as my father put it.

When I feel ready, I tuck into the dive, avert my eyes, and listen for my father's voice. Turning upright for my first ear-clear, the extra weight keeps me sinking the way it should. I pass the top of the coral head, roll into a second ear-clear, and realize I'm not going to make it ten more feet to the declivity with enough breath to do what I need to do. I'll have to do it here and hope for the best.

My father's ashes fill the water in a gray cloud around me as I turn the pillowcase inside out. I can feel them brush against my body as I linger there on my last bit of oxygen, watching a few large fragments of bone spiral down below the cloud into the declivity, where the circling wrasse and baby parrotfish move aside to welcome them.

I'm close enough to the side of the head to be able to reach out and touch the coral that's still alive, as lightly as I can so as not to damage it. Then I have to go.

For a long time after surfacing I float limply, struggling for breath. The gray cloud below me drifts down-tide away from the head, and I imagine I see the silvery glint of a barracuda inside it. Is that possible? Tears are running down my cheeks inside my face mask, the first time I've cried underwater since Susan was lost twelve years before. I cough and sob into my snorkel, but don't want to lift my head and stop looking down. Bursts of sunlight alternate with cloud shadows in a nice rhythm, and the silent world beneath me glows and fades.

Here you are, where you ought to be. The coral might be dying on this, your favorite spot in all the world, but your old friends are still with you. I do hope you're happy.

I hope I've done okay for you.

I hope someone will do the same for me when the time comes.

In this very place.

An image of my father floats into my mind—the one I often see when I think of him. It's not the posed, Charles Atlas one in the front of this book; in fact, I'm not sure where this image comes from or even if it exists outside my imagination. He looks about forty, wearing only his blue nylon swim trunks. He's half standing, knees slightly flexed, braced against the transom of a runabout going at top speed, palms flat on its topside, arms tensed to hold him straight—the runabout's white wake spreads out behind him. His usually neat hair is all over the place, his eyes are slitted, and his teeth show in a wild grin. It's the image of a man I would have loved to have known.

Collecting

Our Faith

November 2005

NOT LONG AFTER OUR FIRST expedition, Dominique is fired in a tsunami of budget cuts and ends up at Millersville University in rural Pennsylvania, where she's swamped with teaching duties and a new baby. Danielle and Heidi are gradually phased out, and I am left with no academy scientist to run the project. *My* project now, I've come to realize.

In desperation, from my loft in New York, I call John Lundberg, the academy's curator of fish. John recommends Loren Kellogg, a forty-one-year-old research associate who has spent the last ten years struggling with an ecologically oriented doctoral thesis on groupers in the Gulf of Mexico.

When I drive down to meet Loren at the academy, I learn some appealing background: he keeps pet groupers ("the most intelligent fish I know") in his aquarium, and as a kid SCUBAed in freshwater swimming holes around Buffalo, New York, where he grew up. He'd given up drinking after a rowdy, rebellious youth—a little like mine—and he knows his Caribbean fish cold.

Over the next few weeks, he eagerly comes up with a first-stage plan for the study that seems to make sense: to select just four of my father's sites that would be representative of the whole group, and to make rotenone collections on them or near them at three different depths to reduce the variables. In the collections, we'd try as much as possible to replicate conditions described in my father's field notes.

One problem: we have yet to formally apply to the Bahamas Department of Fisheries for the rotenone permit. That would have been Dominique's job. Now Loren will handle it, as the project's scientific leader.

We plan a two-man reconnaissance expedition to Nassau, to select the four sites and to meet with Braynen again to tell him our ideas before we make formal application.

Loren looks very much like Jim Böhlke; I realize this when we rendezvous at the Nassau airport. He's overweight, intense, favors baggy Hawaiian shirts and old khaki shorts, and on his feet is a pair of uncompromisingly nerdy Keen sandals, almost as big as snowshoes. Keen sandals hadn't been invented in Jim's day, or he surely would have worn them. I find the Kellogg-Böhlke resemblance both reassuring and unsettling. But I'm ever so happy he's here.

"Weather reports are good," I inform him while he wrestles with his mountain of dive gear, photographic gear, personal library, computer, charts, and GPS.

"Yeah, I've been checking online. There's a little front coming through in a couple of days, but it shouldn't bother us too much." Loren is nothing if not technologically plugged in.

We load his stuff into our rental car and drive to the motel where we're booked, the Orange Hill Beach Inn, conveniently close to Lyford Cay, where I've rented a skiff.

In the late afternoon, after we get settled, I suggest a snorkeling jaunt to Clifton Point, a short drive around the west end of New Providence Island, where the Tongue of the Ocean comes within a hundred yards of land. In the old days, you could stand on the cliffs here and watch huge rainbow parrotfish, barracuda, and reef sharks working the water right below you. Great stands of staghorn coral (the most extensive around New Providence Island) flourished a little farther out, and mako sharks have been spotted here. If the truth be known, I've staged our swim now as a little test. For both of us.

"Want to check out the drop-off?" I ask casually, as we stand on a little beach putting on our gear.

"Sure." Just as casually. Of course he has no idea what we might see, but then at this point in time neither do I.

The staghorn coral has disappeared completely, unless my memory of its location is faulty, and the other shallow corals seem in worse shape than anywhere I've currently seen. No barracudas, no rainbow parrots, no sharks of any kind. As we swim out, the marly bottom slopes gradually down to a depth of forty feet or so and then suddenly ends.

Out here, the visibility is well over one hundred feet. The cliff falls away like the side of the Grand Canyon, in ledges and outcroppings into the dark blue of infinity (6,000 feet—deeper than the Grand Canyon). Coral down there looks fairly healthy, and at least one good-sized black grouper is patrolling the edge. In the sunlit, electric blue of open ocean beyond, a little school of horse-eyed jacks flits hurriedly by like a flock of ducks. The drop-off is a favorite haunt of big pelagic predators like tuna, marlin, dorado, wahoo, and mako and oceanic white-tip sharks, but today they don't appear.

Thirty-eight years ago, across the Tongue of the Ocean off Andros Island, a Canadian diver named Archie Forfar and his partner, Ann Gunderson, died together while trying to set a new compressed air deep-diving record over the drop-off. I've always found their story haunting, but since Susan was lost doubly so.

Forfar ran a little diving operation in Stafford Creek, specializing in wall dives. The wall fascinated him; he never got tired of exploring it. One of his discoveries, Hole in the Wall, is a spectacular underwater shaft, starting at the top of the wall in 120 feet of water and dropping down to 200 feet, where it exits on the vertical cliff. Cousteau took a minisub down the shaft while Forfar was still alive to show him its location, but the secret died with him, and Hole in the Wall was lost for the next twenty years.

The wall has always attracted deep divers, and several previous records on compressed air had been set here: for women, Betty Singer at 312 feet in 1961; and for men, Neal Watson and John Gruener at 437 feet in 1968. With careful decompression, you can avoid the bends after diving to those depths, but two other hazards are unavoidable: nitrogen narcosis ("rapture of the depths") and oxygen toxicity.

Rapture of the depths is controlled by Martini's Law: the effects are compared to those of one martini at 100 feet, and one martini more for each additional 33 feet—a dive to 400 feet would total ten drinks. In its earlier stages, nitrogen narcosis is a great high, much more like laughing gas than alcohol, and when you come up there's no hangover. But severely narc'ed divers are prone to hallucinations, hysteria, terror, and stupefaction and have been known to hand their mouthpieces to passing fish. Oxygen toxicity, for its part, can occur without warning at depths over 200 feet and causes convulsions, nausea, tunnel vision, lung collapse, and retinal detachment.

One curmudgeonly observer in Nassau commented after the Forfar tragedy in a letter to the *Tribune's* editor that diving to record depths on compressed air (rather than using a less toxic oxygen-helium mixture) is comparable to seeking a world's record for arsenic consumption. Still, compressed air is the original SCUBA medium with which all the early records were set, and there are hazards to deep-diving no matter what method you use. It's only the depth that varies. Many people have died trying to set these records, and many authoritative outlets have stopped publishing the results. A published record, for the deep-diving fraternity, can often turn out to be a fatal challenge. In mountain climbing, there's a summit to reach. No one is ever going to SCUBA dive to 6,000 feet and live to tell about it.

Archie Forfar, a lean, hawk-nosed, spade-bearded, thirty-eight-year-old Canadian Scot, ran the dive operation with his wife. Blond, twenty-three-year-old Ann Gunderson, also a Canadian, came as a tourist but stayed on to work for Archie as a dive master. Her fascination for the wall matched his, and she became his partner in exploration.

It became more or less understood that sooner or later the couple was going to go for a male-female joint deep-diving record. Six months before the attempt, they began a series of deeper and deeper dives over the wall to familiarize themselves with the effects of compressed air at great depths and to build up resistance to them—as one might gradually increase one's dose of arsenic, according to the curmudgeonly commentator. Toward the end, they were diving routinely to 350 feet and deeper, where they collected three specimens of the rare Adanson's Slit Shell, the only ones ever obtained on compressed air.

To beat Watson and Gruener's record, they planned an 11.5 martini dive to 450 feet. Down there, light would have dwindled to less than 1 percent of surface intensity, and the water temperature would have dropped into the forties. They'd wear thick wet suits, carry underwater lights, and use two tanks apiece. Other tanks to use for the decompression process would be tied off at strategic intervals.

Three support divers would wait at depths of 10, 150, and 300 feet, while another would accompany them as far as he could. Archie, Ann, and the accompanying diver, carrying ten-pound weights to help them down, would descend a line anchored to a ledge at 500 feet. When they reached their goal—five minutes or so for the descent—they'd jettison their weights, attach a clip to the line to prove they'd been there, and start back up to the first decompression point. Decompression would take three hours. According to their close friend Pastor Robert Raburn, who'd been a guest on their dive boat, the couple agreed that if one of them ran into difficulties at depth, under no circumstances would the other risk his or her life trying to help.

I obtained Pastor Raburn's account of what happened on December 11, 1971, given to a reporter for the Nassau *Tribune* and published on December 18, through the US Library of Congress. It appeared under the headline: *DID FATAL DIVE PAIR BREAK THEIR "NO-HELP" PACT IN RECORD BID?*

> The first indication that something was wrong came when one of the men from the support boat surfaced half an hour later (the divers had entered the water at 10 AM) and told the two Commissioners that he had gone down to 150 feet and could see only four divers on the line.
>
> At first it could not be determined which two were missing until the diver went back down with a slate and got a written signal that it was Archie and Ann.
>
> Because it takes a diver three hours to decompress before surfacing, the spectators had no inkling of what had transpired underwater until 1:30 PM when Jim Lockwood came up. Two other divers followed at 2 PM and the third at 2:30 PM.
>
> As Jim told the story, he, Ann, and Archie had reached a depth of 400 feet according to his gauge when he felt himself passing out and inflated his vest. The rest of the story came from the other divers.
>
> Jim had blacked out and started rising swiftly to the surface but was caught at 250 feet by Randy. [The pastor did not know Randy's last name.]
>
> Randy took Lockwood to one of the suspended air tanks and held him there until he straightened out. Thinking he was conscious, he turned him loose only to see him start shooting up again, indicating he was still unconscious.
>
> Fortunately he was caught at 80 feet by the other safety diver who took him down to 150 feet and held him until he was completely clear and had regained consciousness.

As Lockwood surfaced, two of the other divers started down to look for Archie and Ann. The first was Schick, the deep-safety man, followed by Randy.

At 450 feet, Randy got tunnel vision (a symptom of oxygen toxicity in which the field of vision narrows to a small point). Schick meantime had reached 470 feet and could see Ann and Archie, their flippers moving and exhausting oxygen. He could tell, however, that something was wrong.

Ann was ten feet away from the marker line, lying on the ledge, and Archie was leaning over the line's 300-pound anchor weight.

At this point Schick too began to experience tunnel vision. When his vision cleared and he saw them again, both were still and no oxygen was being emitted.

He realized he couldn't stay at that depth and that he couldn't reach them because they were some 25 to 30 feet below him. So he started back up, reaching the surface at 2:30.

At 2:30 PM what was left of the group was picked up and was taken to Fresh Creek where they made their statements to the police corporal there.

Not one of the divers was even prepared to hazard a guess as to what could have gone wrong. Only the day before, Archie and Ann had dived to 475 feet, and the week before they had gone to 450 feet, the record they were aiming for.

They had taken every precaution, but perhaps instinct is greater than reason.

As Pastor Raburn put it: "One of them could have run into trouble, and the other, unmindful of the promise made, went to the rescue. At that depth, any exertion spelled death for both."

Coda: Not long after Archie's and Ann's deaths, two of the three rare slit shells they'd collected at depth were stolen, according to a shell expert who had studied them. He reported in a shell collecting journal that "the remaining specimen was kept by Mrs. Archie Forfar."

I always assumed, maybe wrongly, that their love of the deep blue was only one aspect of the partnership between Archie and Ann, that setting a male-female dive record also would have been a consummation of their love for each other. Under the circumstances, the "no-help" pact was ridiculous unless they both planned to stay down there forever.

When Super Typhoon Gay was taking aim at our little sailboat on November 18, 1992, Susan and I were anchored at Wotho Atoll in the Marshall Islands, halfway across the Pacific. According to our weatherfax and single-sideband weather reports, the typhoon most likely would pass north of us sometime in the middle of the night. We elected to stay with our ship. Meanwhile, unbeknownst to us, Typhoon Gay swerved twenty degrees southward. At about one a.m., the outer edge of the eye had reached us. The eye itself was to pass right over us.

In 105-knot winds, our little sailboat was torn from her anchors, blown over on her side across the atoll's lagoon, and smashed apart on the encircling reef. At two a.m. Susan and I, wearing inflatable life vests and tied together with a ten-foot nylon line, found ourselves rolling helplessly in heavy surf breaking across the reef. The element we breathed was spume: half water, half air. The darkness was absolute.

First we came out of our life vests, then we came out of the fixed loops in our safety line. I was behind Susan, holding her around the waist with one arm and holding on to our remaining life vest with the other.

"Hold me up," I heard her call, then, "Oh, oh, oh," before another wave hurled us down and tumbled us underwater.

Then she was out of my grip; I felt her long hair brush my fingertips on her way down. My choice was to let go of the life vest and try to swim down after her . . . or not. Unlike Archie Forfar, I'll have to live with what I chose.

I can feel the lure of the deep blue now, with Loren, even if I'm only floating on the surface above it . . . the urge to keep going down, farther and farther, until you find what you're looking for.

My new wife, Sarah, and I are happy, even blissful, yet I don't think I'll ever overcome the seductive pull I experienced alone in the typhoon after Susan had gone, in the same total darkness as I'd find down there over the drop-off, waiting to go. *Wanting* to go. Wishing I had gone.

I reentered the ocean for the first time after Susan's death on a beach in Baja California, near the little town where we'd spent a year in the early eighties. It was Christmastime, a month after the shipwreck. I'd come because the huge ocean beach had been one of her favorite places.

Early morning. I was alone. The beach was deserted, south to some distant cliffs and north unbroken for more than a hundred miles. The air was luminous with sunlit spume from the fifteen-foot waves rising up out of the deep water and exploding on the sand. The bombardment had woken me in my bed a mile inland, and I could feel the house's thick old adobe walls shake as I lay there in the dark.

There was no wind, and the ruler-straight line of each swell as it marched toward the beach was visible for at least a quarter mile out. Beyond that, the shiny black shape of a breaching humpback whale flew high into the air and splashed back down in a white explosion. Brown pelicans surfed the wind currents on the wave faces, rising up as they broke then dropping down to continue on the next. Caught in the transparent faces, silver sierra mackerel chased clouds of silver minnows. The air was tangy and salty as an oyster.

Early on in our time together, Susan had sent me a card illustrated Frank Frazetta–style with a dark, helmeted archer high on the crest of a wave pulling a sea nymph from a bubble to kiss her. She'd written in the margin:

"Gordon, my love, isn't this card wonderfully symbolic?"

So I knew I was going in. I'd known it when I first heard the waves.

I waited for a lull, ran into the thick, swirling soup, and started stroking out as fast as I could. A couple of smaller waves broke in front of me; I ducked under them and kept swimming. Then the first wave of the next big set began to rise.

I could have tried diving for the bottom. Instead, I stroked straight up the face, higher and higher as it lifted me close to the vertical, next door to heaven. I waited for the breaking curl to grab me, hurl me back on the sand, and crush me.

But at the critical moment, my head and upper body emerging into thin air like the archer in Susan's card, the force relaxed, I felt a light slap on my face almost like a caress, and the curl passed below to explode on the beach without me.

Out beyond the break, the big smooth swells passing dreamily beneath me, I caught my breath, rolled into a dive, and stroked my way down as far as I could. The familiar blue vastness of the ocean was all I could see. She was out there. And, somehow, she got me safely back to shore.

I sense movement in the water next to me and look around quickly. Loren has jackknifed into a dive, hands sweeping back to his sides, big black fins flapping slowly and rhythmically. At about twenty feet I can see him grab his nose to blow and clear his ears. He's one of the lucky ones . . . *Eustachian tubes like Grand Prix exhausts!*

Just above the lip he stops and hangs there like a stationary skydiver, outlined sharply against the deep blue. I can see him check the depth gauge on his wrist. I count: One. Two. Three. Four. Five . . . With a quick hand motion, he's turned himself upright and is unhurriedly soaring back in my direction. On the surface he pops his head up and takes out his snorkel to talk. I join him. He's smiling broadly. "Did you see the big yellowmouth?"

"Yellowmouth?"

"Yellowmouth grouper. Just over the edge."

"No, I couldn't see him." I don't even know what a yellowmouth grouper looks like. "How far down did you get?"

"Forty-three feet."

I'm impressed. Well, he's quite a bit younger than me. "Coral down there looks in pretty good shape, doesn't it?"

"Yes it does." Somehow he manages to float and talk with no extra effort. "This place reaffirms my faith."

I couldn't have put it better. We pull our masks back over our faces, refit our snorkels into our mouths, and stare down into the blue. The sun comes out from behind a little cloud, and shafts of light slant down through all that misty space like they do in the vast Cathedral of Saint John the Divine, back in New York City.

The Boat Who Wouldn't Float

July 2006

EIGHT MONTHS AFTER LOREN AND I made our recon trip and two years after Dominique started things off, the Chaplin Project's first actual collecting expedition begins with an uncomfortable feeling that it's all too good to be true.

Director of Fisheries Michael Braynen has done us proud: not only do we get the rotenone permit (after Loren forwarded a compendium of scholarly papers showing that the organic root extract breaks down rapidly after use, and that an exposed area completely recovers in four to twelve weeks), we also get use of the Fisheries Department's sixty-five-foot trawler *Guanahani*, her captain, and two crew. Completely free of charge for an entire week, including an air compressor for SCUBA tanks, and an eighteen-foot outboard runabout. All we have to do is pay for the fuel and take some Fisheries interns with us.

Where's the catch? My favorite nautical narrative is Farley Mowat's *The Boat Who Wouldn't Float,* a sequence of misadventures that make such great reading you don't even have to believe they actually happened. I am to think of this book many times during this trip.

Loren has put together a multidisciplinary team: Dr. Walter C. Jaap, a choleric, seventyish coral expert from the University of South Florida; his pretty blond mid-twentyish assistant, Jen Dupont; Heidi Hertler, who has taken two weeks off from her new post at the InterAmerican University in Puerto Rico; and me. We aim to make rotenone collections on the four Böhlke-Chaplin sites we selected on our previous trip, plus coral surveys to establish exactly what the cover is now and how it might have changed from the old days. As before, we're backed by funds I have raised and by my own contributions.

We arrive to find a tropical trough has settled over Nassau: overcast skies, thunderstorms, and squalls. With a wide smile but indirect eyes, Captain Rollie takes us to inspect the *Guanahani*, lying at Potter's Cay under the bridge to Paradise Island along with the remnants of Nassau's fishing fleet and out-island bumboats. The old trawler looks long unused, stained, and rusty, secured to the pier with steel cables instead of dock lines. On the foredeck beside the anchor windlass is a large puddle of hydraulic fluid. We'll need to buy several gallons of it in the morning, Rollie says, because the leak is bad.

Rollie unlocks the padlock on the main cabin door and our team files inside. He wishes he had known about our trip in advance—then he could have gotten the refrigerator, the toilet, the sink, and the freshwater wash-down working properly, and he could have rebuilt the dive ladder. In the musty gloom of the galley/stateroom, Heidi and I look at each other and roll our eyes. . . . *When the going gets tough, the tough get going.* Thank God she's here.

The compressor is in a wooden shipping crate in a nearby storeroom. It's brand-new, has never been unpacked. Who knows whether we can get it to function? Even if we can, it's not big enough to charge more than a few tanks daily. We pile back into our rental van and head to the dive shop to arrange equipment rentals for the next morning, including twenty-five SCUBA tanks that we'll have

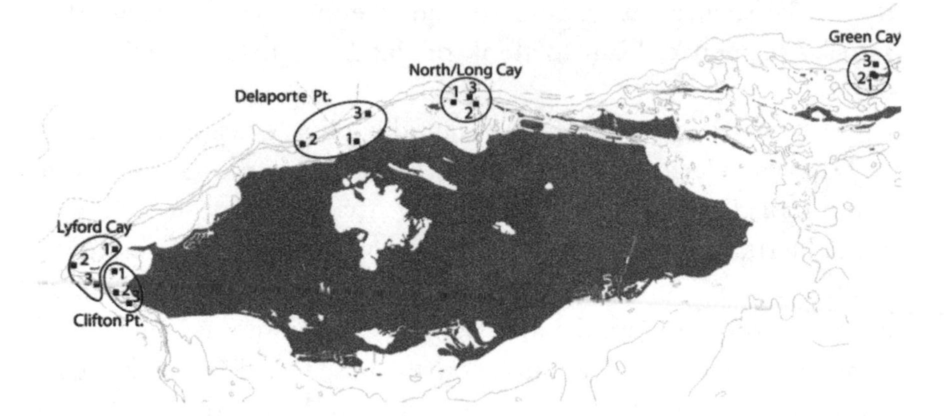

to wrestle daily from the shop to the van to the boat and back again for the recharge.

Thunder squalls with heavy rain and howling winds keep us in port the next day—all twelve of us. Besides Loren's scientific team, our number has grown to include Rollie and three crewmen; dread-locked Kogi, our liaison with the Fisheries Department; and two uniformed Royal Bahamas Defence Force officers complete with semiautomatic weapons, who for some unknown reason have been mandated by Fisheries.

So Loren has plenty of time to outline his complex strategy for the four sites. Before making the actual rotenone collections, we will measure the reef's profile (its varying height from the bottom) in three separate transects, survey the coral cover, conduct visual fish counts, and film video along the transects. This will be so time-consuming that some of us estimate we will only be able to complete half the stations in the days left to us. But Loren will not compromise. All surveys must be done, even if this means another trip.

There are not enough lead weights on board, and some of the underwater pencils don't write properly. Tempers are running a little short, and when a RBDF officer announces we can't use the outboard runabout because the fuel gauge doesn't work, some of us lose it.

"I'm just trying to help you," he shrugs.

Walter Jaap is shaking his head, his face red, and a vein is beating in his temple. "When things start to spin out of control is when they get interesting," Loren observes. We finally resolve the runabout's fuel gauge issue with a dipstick.

The next day, in spite of a strong southeast breeze and a late start (Rollie forgot the keys to the anchor lazaret), we're able to make one collection station close to the harbor. It's in twenty-five feet of water beyond the reef connecting two little cays west of Paradise Island. We came here often in the old days to collect in the shallower caves. We use two gallons of undiluted rotenone, applied by Loren and me from polyethylene squeeze bottles, and collect a grand total of seventeen fish, where my father and Jim Böhlke used to get hundreds.

"I feel kind of let down," Loren says.

For some reason, it doesn't occur to us that we're doing it wrong. Maybe the disappointing collection was just bad luck. We try again the day after on the shallow and mid-depth collection levels farther afield near Green Cay. The dark brown liquid issues from the bottles in oily round blobs, not mixing with the seawater but coating our exposed skin. I'm wearing only a wet-suit top instead of a protective full suit like the others; by the end of the day my hands are tingling and my penis is completely numb, while the rest of the team show no effects. In the tiny, foul-smelling head (which we've gotten to evacuate but not to flush), as the *Guanahani* chugs slowly home, I desperately try to massage it back to life with no success. I feel dizzy and my vision blurs.

Meanwhile, most of the fish we have been trying to collect remain uncollected. A handsome Fisheries intern named Ashley (daughter of the Minister of Agriculture), with tipped hair and a Valley Girl accent, who's been with us for the last couple of days, nets the few specimens we get and passes them to Wellie, our thin, hyperactive launch pilot who has enough good spirits for all of us:

"You gots to learn something new every day! Dass what life all about!" Is he high on life? "Dese fish must be *berry* important, dass all I can say."

Stately summer cumulous clouds drift from east to west, chasing their shadows on the glassy water, as the *Guanahani* snakes her way out to Green Cay the following morning for the deep rotenone collection. The weather has finally given us a break; even the fact that Rollie can't seem to steer in a straight line is endearing. William, the portly half of our Mutt and Jeff crew, mimes a toke and laughs. "Wellie flyin' with the an-*gels* dis mawnin."

"I hear what you said," Wellie grins back.

I'm flying with the angels myself: below the waist, I'm almost back to normal.

I've not told Loren that we'll be making this collection on my father's final resting place; I want to keep that slightly daunting knowledge to myself. But I do think the old blade would love the idea and pray he'll do everything in his power to make it a success after our lackluster string of failures.

This time we're going to mix the rotenone with seawater in a one to ten proportion before we dispense it on the reef. How did we do it fifty years ago? I can only remember emulsifying the powder with seawater in the early days, and though I know we graduated later to the liquid form I have no idea how it was mixed. Loren has combed the field notes in vain for references. Why didn't we consult outside sources before this trip? Scientists are often proud and secretive, that's why.

Loren's faithful GPS finds the deep head right on the button (he already entered the coordinates during our two-man reconnaissance trip). This takes away some of the suspense, some of the excitement of rediscovery, but we have a full boat—our five-person scientific team, Captain Rollie, crewmen Wellie and William, a single armed RBDF officer, a Fisheries liaison, and Ashley the handsome intern. Time is of the essence.

For the first time in fifty years, I'm SCUBA-soaring around the familiar promontories of my father's favorite coral castle, now his memorial. I'm confident our collection here is going to be memorable. How can it not be?

Most of the coral is dead and covered with algae, but, as I saw when I deposited his ashes, the towering, crenellated shape of the reef hasn't changed. I still feel I'm about to fall, sailing over the edge of a drop-off. The gloom of the caverns still electrifies with possibilities. The big groupers that used to lurk there are no longer around; neither are the big snappers or the free-swimming silvery jacks, mackerel, and barracudas. But all our old friends that inhabit the reef itself—the tiny, brilliant basslets, chromis, damselfish, wrasse, parrotfish, cardinalfish, et cetera—seem still to be there in about the same numbers and diversity as I remember. Or is it my imagination?

We have about twenty gallons of the rotenone-seawater mix to put down. Loren, Heidi, Jen Dupont, and I carry either a collapsible five-gallon plastic jug or a double-thick heavy-duty trash bag lashed shut around a polyethylene tube. We position ourselves east, west, north, and south, start at the bottom, and work our way slowly up the tower, squeezing out mixture as we go. There are still a fair number of unemulsified blobs in it, but they are smaller than before.

Hanging in mid-water, waiting to see how well the mixture will work, we are joined by Kogi, the Fisheries liaison, Ashley the intern, and Walter Jaap. They distribute long-bellied collecting nets, and I pray to my father that there will be something to put in them.

He seems so close that our meager collection is especially heartbreaking, about fifty times smaller than those he'd made here fifty years ago. Loren tries to look on the bright side: if the rotenone tests out at standard efficacy, a 50 percent reduction of the fish population over time can be extrapolated . . . A major finding. But my own eyes tell a different story. I can see that the fish are there.

"What's the next phase?" Walter Jaap asks when we are all back aboard the *Guanahani*, that vein beating in his temple.

Ashore, my eyes begin to sting ominously, probably from the rotenone oils, and I watch through a veil of tears the population of Nassau joyously preparing for Independence Day. There is dancing all night in the streets, while I toss and turn in the motel room I share with Loren, and he stays up until the wee hours combing through my father's field notes for mixing hints.

I wake to a gray fog. Rotenone irritation makes the insides of my eyelids feel like sandpaper. "Listen . . . I'm afraid I can't see very well," I tell Loren.

A man for all seasons, he's brought antibiotic eyedrops. I put them in and hope for the best, but stay ashore with my eyes closed all day. The fog has mostly lifted by the time the team returns, but collection results remain dismal.

At two thirty a.m. the next day (our next-to-last), Loren comes across the word "buffer" in my father's field notes. By the time I wake up, he's translated the word to "detergent," and I lie there wondering why I hadn't been able to remember this from the old days. And wondering why I myself hadn't combed my father's field notes more thoroughly before the trip. Detergent should neutralize the oily base of the liquid rotenone and allow it to mix with water.

The new recipe is: 3 cups Joy Ultra Concentrated detergent, 3 gallons Chemfish Regular 5 percent rotenone, 15 gallons seawater. Will it work this time? The *Guanahani* snakes out of Lyford Cay harbor without mishap (it had sheared off a channel marker entering the day before) and heads east toward Nassau, anchoring for a deep collection in forty-five feet of water beyond that same shallow reef connecting the two small cays near Paradise Island where we made our first try. The vessel is listing sharply to port: "The water versus the fuel," explains Rollie.

The weather's good, a light southwest wind, no squalls. Another in a series of Fisheries interns, who spent the day before seasick on the settee, is talking nonstop on her cell phone. Kogi arranges for Wellie to pick up a lunch of Kentucky Fried Chicken onshore, via the launch. "See? We Bahamians just copy everything you do," he says, grinning at us.

This is a perfect site: an isolated coral head, mostly dead but with a good population of fish, including seven large blue parrotfish and a three-foot tiger grouper. We make the coral cover estimates, visual fish counts, transects, and reef profiles and board the *Guanahani* again to mix the rotenone.

I watch Loren's face as he measures out the cup of amber Joy soap. He feels my eyes, looks up, and raises his eyebrows. Kismet. We pass the two trash bags of soap and rotenone mixture down to Wellie in the launch, and Loren climbs down with him to hold the bags while Wellie pours in water from a five-gallon bucket. I don my SCUBA gear and jump over the *Guanahani's* side to receive one of the bags. Loren enters the water from the launch with the other, and we start down toward opposite sides of the head.

As soon as the thick white mixture starts spouting from my tube, I know we've finally got it right. The rotenone is now mixing with the seawater instead of floating around in oily blobs. In the end, we collect over two hundred specimens, including many old friends such as the orange-and-red-striped peppermint basslet, the predatory wrasse blenny disguised as a harmless juvenile bluehead, the trout-like spotted soapfish, and the burrowing mottled jawfish, which carries its eggs in its mouth.

Aboard the *Guanahani*, we set up a sorting table where the specimens can be examined and tissue samples taken, labeled, and put into tiny vials of alcohol for future DNA testing. (In my father's day, this didn't exist.) Loren, Heidi, Ashley, and Jen do the work in a festive atmosphere.

"I get excited when things work," says Heidi. She wants to get her graduate students involved: "There are a lot of MAs in this data."

There is talk of presenting the results at conferences, raising funds, publishing peer-reviewed papers in prestigious journals. Even Walter Jaap seems galvanized and happy.

The atmosphere takes me back to the old days of simple collecting, the excitement of discovering new species, where everyone's eyes were glowing as we sorted through the specimens. There are no new species here (at least not at first glance), but there is the possibility of something more exciting being turned up when our new collections are rigorously compared to the old ones back in the academy labs: proof that the original species are still around.

We make another successful collection on our last day, leaving ten to go. In other words, we don't even come close to finishing the survey on this trip. We'll have to raise more funds, get another team together, and come down one more time.

A Woman's Point of View

July 2006

After the expedition, my wife, Sarah, flies down to Nassau from New York to spend the weekend before we go back together, and once again I thank my lucky stars that the effects of rotenone on human male flesh don't seem to be permanent. My vision is almost back to normal, too.

Sarah has never seen the Chaplin House, and it's been two years since I've seen it myself. So, on Saturday morning we drop the rest of the team off at the duty-free shops on Bay Street and drive the expedition van across the Paradise Island bridge to Atlantis.

I've been advised we can avoid unpleasantness on the beach approach to the Chaplin House by driving through a mini Holland Tunnel under the canal system, through the construction zones, to the Club Med boundary where we can park and walk through a gate. From there it's a hassle-free walk.

There's a guard at the gate, of course. He wants to know our names. Then he consults a list. "Sorry. You not on here."

"What do you mean?" Sarah has red hair and a short temper.

"This a list of *res*-idents up there. You not on it." He fingers his VHF. In a minute he's going to call for backup. Perfect! My earlier wish is going to come true: the two of us are going to be arrested. Arrested trying to go home. Maybe it will make the papers!

But then he begins to fidget a little, and finally he opens the gate. "Aw, go ahead, mahn. But don't tell anyone I let you through, arigh'? These folks *particular*."

The big wooden sign advertising the Chaplin House is still out on the beach, and the house itself looks the same, but there's no one around. We peer in windows to see that most of the rooms have been stripped of furniture.

I remember how Ronnie had been complaining that he was sick of Nassau, wanted to sell the place and move to an Out Island. But I was stunned that it had actually happened. "I never thought he'd do it," I tell Sarah. "He loved the place. He'd been here for thirty *years*."

I show her the spot on the north verandah where my father taught me to read, and his soundproofed bedroom with its west and north views of the ocean. When we walk through the connecting door to my mother's room overlooking the garden, Sarah shakes her head. "But it's so stuffy and closed in. His is much nicer."

"She always had the air conditioning on. She didn't want open windows."

"No open windows? Here in this wonderful place?"

"No. She didn't go outside that often. She got terrible sunburn."

"Well, what *did* she do?"

"Had friends down from Philadelphia. Gave dinner parties. She did go on picnics, if she could stay in the shade. She had a parasol. . . ." Trying to describe my mother's life to someone who hadn't known her was almost impossible. "She, ah . . ."

"Was she interested in fish? Did she snorkel or dive?"

"Jesus no."

"So. She was here just for your father's sake? Didn't she *pay* for it all."

Sarah's blue-eyed intensity is making me uncomfortable, even defensive. "Look. She had some good times here, in the early days. It was just later . . ."

In 1961 my mother fell on a trick knee during a game of charades and became a semi-invalid. No doctor she saw was able to give an exact diagnosis, and the sound of her icy voice issuing orders from her bedroom became the sound of home for me. "Louise is withdrawing more and more into a cocoon, or rather an igloo of self-pity," my father wrote in his diary.

In Philadelphia she canceled two operations that the doctors said would probably bring relief by fusing the knee. Why? I secretly thought she might be enjoying at least one part of her invalidism—the part where she was in absolute control, with my father waiting on her hand and foot.

Finally, my father talked her into seeing a psychiatrist. "Spent an hour with Dr. Shefelen," his diary reads, "during which time he tried to explain his interpretation of Louise's complicated psychic make-up, which seems like something from a Dickens story."

I would give a lot to have heard Dr. Shefelen's interpretation, though it's not hard to imagine the basic outline. As a child, in addition to her sadistic German-Swiss governess and her distant club-footed father, my mother had had to deal with her own mother's invalidism.

That started with the birth of my uncle Cummins and lasted for the next twenty years until her death from breast cancer. No doctor could come up with a diagnosis in her case either. The venue was very significant: the same bedroom in the gloomy Philadelphia house that my mother occupied when she wasn't in Nassau.

So I say to Sarah exactly what my mother's old friends and relatives always said to me: "You have to remember she had an unhappy childhood."

My wife is peering out the small, louvered, almost opaque window into the garden. In some ways, she is like my mother was before her fall: impulsive and adventurous, not afraid to pursue the unconventional (marrying a man who's twenty years older, while my mother married a penniless Englishman).

In a matter-of-fact voice, she asks a question that I have been sparring with since my mother died in an automobile accident in the fall of 1983. "Do you think she drove in front of that truck on purpose?"

I was notified of the accident by a phone call from my father's secretary and arrived at my parents' house in Haverford, on Philadelphia's Main Line, at about seven thirty p.m., September 19, 1983. "Your father's out to dinner," said Grace, the maid. "But Hannah's cooked you some chops and fresh peas and your favorite brown Betty for dessert."

At the hospital, a nurse said my mother was too tired for visitors, and to come out the next morning.

After dinner, I wandered into the living room and sat at my father's big red leather-topped partners desk. The prodigal son. In the carved mahogany tray where he kept incoming mail was a gossipy letter from his old Nassau diving buddy Stan Waterman, who had gone on to become a famous underwater cameraman, expedition leader, and lecturer. Stan was planning a trip to the Cocos Keeling Islands, in the middle of the Indian Ocean, the private domain of a Scottish laird whose ancestors had been granted the land hundreds of years earlier. I would have given my left arm to go on this trip, but my father had never pulled any strings with Stan on my behalf. (Many years later, visiting Stan in Maine, I was to learn my father had given him not only his beautiful Boss 12-gauge shotgun, the kind of thing you pass on to your son, but also the little .410 double barrel he'd originally given me as a boy. Stan was embarrassed to note my initials were still on the case and offered to give it back to me. I proudly declined.)

It suddenly occurred to me, sitting at his desk, that my father had washed his hands of me. I was a loose cannon, an adulterous one at that. I'd quit a perfectly good job in journalism to try to make it as a freelance writer and novelist. I'd abandoned my two teenage daughters and my wife of thirteen years to run off with another woman (at the time, Susan and I were renting a house in Orient, Long Island). All I did these days was drive my mother crazy with more and more demands for money. She and Diana, my older daughter, would whisper in her room for hours about how horrible I was.

I was in bed at one a.m., when I heard his car in the driveway. I didn't bother to get up.

I forget exactly what my father and I said to each other the next day, driving out to the hospital, but the crash had been much worse than I'd been led to believe. My mother's Saab had been hit broadside by a speeding eighteen-wheeler and totaled. She was in intensive care with four broken ribs, a broken collarbone and shoulder blade, and many cuts and contusions. The fact that he hadn't bothered to call to let me know that she was in critical condition told me exactly where I stood. Low priority.

In the hospital, they said she'd taken a turn for the worse and was now in a coma, on life support. We stood at her bedside and watched the green line on the EKG monitor record her heartbeat. When the doctor came in and shone a light into her open eyes, the pupils did not change size. "They look awfully dry," my father stammered, and the doctor absently squirted in a few drops of solution.

"She wanted me to go to the dinner party," he explained without my asking. "She was dying to hear the gossip." This irony passed him by. "Before I left, I asked what else I could do for her, and she said 'nothing.' That was the last thing I heard her say."

He didn't look at me. I wanted to reach out and put my arm around him but hadn't done anything like that since I was a child. I was aware he'd been through some kind of hell with my mother for a long time, but we'd never discussed that. Nor did we ponder her black, possibly bipolar, moods. This was the English way: stiff upper lip. Life goes on.

My sister, Susie, flew in that evening, and we spent the night in a nearby motel. Just before dawn, the hospital called and we got there in time to see the EKG monitor go flat. That sight caused my father to burst into tears, the only time I'd ever seen him cry.

So now, in my mother's stripped bedroom in the Chaplin House, the question was on the table. Had she killed herself?

"Nobody else in the family thinks so," I answer Sarah. "Or if they do, they're not telling me."

"What do *you* think?"

"I think she did."

There. I've said it. How many times in my mind have I been in my mother's Saab with her, while she waited to pull out of the little side road to cross the four-lane highway, seeing the big car-carrier truck bearing down at seventy-five miles an hour and just stepping on the gas? To me, it was completely in character: *you have to act fast when you make a decision.* Sometimes, since Susan died, I could see myself doing the same thing.

"Look. She apologized to at least one bystander at the scene, saying it was her fault. And before she died at the hospital, she refused all medical treatment. A transfusion could have saved her life. Her doctor said so in a deposition. But she said she was afraid of AIDS. This was 1983 . . . AIDS had hardly been invented." I stop for breath.

"The poor thing," Sarah says. Outside, we can see a young woman walking up the path.

"You know, we sued the fucking truck company for reckless driving, but . . ."

A young woman climbs the steps to the verandah, hears me talking, and calls: "Hello? Who's there?"

She's the fiancée of the son of the Chaplin House's new owner. The new owner is David Kosoy, a Canadian shopping mall developer who, a few years ago, bought the old Huntington Hartford place next door from the estate of actor Richard Harris.

The young woman's intended shows up as we stand there talking and seems politely impressed that I'm one of the Chaplins of the Chaplin House. I ask him what his father's plans are for it.

His smile is bland as a spoonful of blancmange. "We'll be tearing it down for condo construction, I'm pretty sure."

Miggy

June 1991

SHE WAS THE REAL REASON behind my mother's decision to get rid of the Chaplin House. Not sunburn. Not boredom. Not depression.

Another woman.

I discovered her in 1991, after my father died: a heavy manila envelope labeled "PLEASE BURN WITHOUT OPENING" far back in one of his desk drawers. It was kind of a joke, wasn't it? He must have wanted it opened, duly recorded, but taken with a certain amusement. Like he himself took all those articles in the *National Enquirer. They found a werewolf in the Maine woods . . . Most amazing!*

Of course, they were love letters, from a woman I'd never met, never even heard of. They were dated from 1962 to 1972, and the last two letters were tragic. In my father's diary for that year, on November 27, he'd pasted a death notice from the *New York Times*:

> WILLAUMEZ, Countess Margaret, died at her home at Capri on Nov. 27. She was the daughter of the late Col.

Fitzhugh Minnigerode. She is survived by her mother, Mrs. Patricia Minnigerode, of Florence. Interment in Capri.

His discreet diary entry reads: "Miggy died today in Capri, poor girl, she had a terrible period of misery and suffering from last June to now, with the same thing that Catherine had [colon cancer]. She took an overdose of sleeping pills in September but survived only to have to go through the trials of a colostomy. Too late." Not many clues there. He could have been writing about a neighbor.

Her letters were different. The first one, dated March 16, 1962 (six months after my mother's fall), and written on someone else's embossed stationery with an Upper East Side address, began:

> Charlie dear,
>
> Thank you for your letter which, as always, was a delight. Yes, darling, you got the goose message. I thought you might be considering straying from the straight and narrow (as they say) and naturally that's why I lent you the book.
>
> Actually I think it must be more difficult for a gander to be faithful than for us humans; to a gander any goose must look good—be that as it may I love you and hope to catch a glimpse of you soon again…

My father must have responded positively and lastingly to this sweet invitation, because nine years later, on a Pan Am flight from Miami to Rome, Miggy wrote: "We flew right over Nassau and I could plainly make out Hog Is., the lighthouse, the curve of Paradise Beach, and your house way down there—also the islands we snooped together, and it made my heart ache."

The next year she was dead.

I sat there numbly at my father's desk, the blue embossed stationery piled on the glass top in front of me. Under the glass were two snatches of verse I'd read many times.

One from John Donne:

No man is an *Iland*,
Intire of it selfe;
Every man is a peece of the *Continent*,
A part of the *maine*;
If a Clod bee washed away by the *Sea*,
Europe is the lesse,
As well as if a *Promontorie* were,
As well as if a *Mannor* of thy *friends*
Or of *thine own* were;
Any mans *death* diminishes *me*,
Because I am involved in *Mankinde*;
And therefore never send to know
For whom the *bell* tolls;
It tolls for *thee*.

The other from Robert Frost:

The sun was warm but the wind was chill.
You know how it is with an April day
When the sun is out and the wind is still,
You're one month on in the middle of May.
But if you so much as dare to speak,
A cloud comes over the sunlit arch,
A wind comes off a frozen peak,
And you're two months back in the middle of March.

Passion and scandal didn't fit at all with those familiar, respectable passages, but there the letters lay: pieces I'd never suspected of the charming puzzle that was my father. Did they smell of perfume?

Chanel No. 5, I was almost positive. It made me shed tears . . . for my father and his mistress and for myself: I'd been left in the dark about this crucial part of his life. I didn't even know what Miggy looked like.

From her letters I was able to glean that she'd been tall, had worn minis in the late sixties, had a villa in Capri, had many friends in New York with Upper East Side addresses with whom she stayed on visits, was part of a jet-setty European crowd complete with yachts in the Mediterranean and chalets in Switzerland, but had little money of her own.

The Internet came in handy. She was the youngest of three handsome daughters of Lieutenant Colonel Fitzhugh Lee Minnigerode who had retired from the US Army to become a military correspondent for the *New York Times*. On October 25, 1941, when she was probably in her early twenties and working as a fashion editor at *Harper's Bazaar*, she married Count René Bouët-Willaumez in Reno, Nevada. The same day, Count Willaumez, forty-one, had divorced his previous wife, Isabel, with whom he had lived a glittery life in New York while pursuing a highly successful career as a fashion artist for magazines like *Vogue*. Presumably, by the time she met my father, Miggy and the Count were no longer one, though she kept his name and title. He died in 1979, seven years after she did.

Our family scrapbooks produced no designated image of Miggy, but there was one uncaptioned photo that I thought was a likely candidate. She's standing in profile on the terrace of a villa, the tiled roofs of the town below sloping down to a distant bay with mountains on the other side. She is pointing out something with a long arm; her forefinger is angled slightly up. She's wearing a blue Pucci dress, three strings of pearls, a thin silver bracelet, and large, square aquamarine earrings, and her dark brown hair is bobbed around her widow's peak. Her nose and chin jut regally out toward the horizon, her mouth is slightly open, and her thin lips are red, but she is not smiling. She looks as if she's making a pronouncement. She's not as pretty as I would have thought.

"PLEASE BURN WITHOUT OPENING." *Duh-hunh.* I can almost hear the laugh. But my father must have wanted me to know he'd conducted the affair discreetly and honorably and had elected to stay by my mother's side in her time of need. If Miggy hadn't died, would he eventually have gone with her? Love is hard to pin down, but at least one concrete and practical aspect of the situation makes me think not. Without my mother's backing, *Fishes of the Bahamas* would never have seen the light of day, and her bequest was continuing to fund the Chaplin Chair, now occupied by Jim Böhlke.

My second cousin, Charlie Grimes, who was handling our family accounts and legal matters, eventually told me how my mother had discovered the affair. While Miggy was still living, one of those incriminating letters had dropped from a book in my father's bedroom in the Chaplin House, through which my mother just happened to be leafing. It was exactly the way my own wife discovered my affair with Susan.

When my mother decreed the Chaplin House must go, Charlie Grimes had proposed the transfer scheme: you don't have to sell, just give it in tax-exempt installments to your son and daughter. That way you'll have access without ownership, and they'll have the nest eggs they've been clamoring for. Of course they're not in a position to take the place on financially, but you'll help them, won't you?

Amazingly enough, my mother was okay with this idea. It would have worked perfectly if I hadn't been banned from the Bahamas, apparently for life, right when it was being formulated. My sister didn't want to take on the place alone.

The Saddle Squirrelfish

November 2006

NOON. OVERCAST, FLAT CALM, DELICIOUSLY humid. Loren, Heidi, and I are back in beautiful Nassau after a five-month hiatus, staying at a run-down spa on the waterfront east of Potter's Cay, where the good old *Guanahani* is docked. We have nine days to make ten collections. If we succeed, we're finished with the collecting phase of the Chaplin Project. Then the lab work can begin, and finally a scientific paper will be written and published in some prestigious journal. We'll all be famous. There is light at the end of the tunnel.

Not so fast, as Farley Mowat might have put it in *The Boat Who Wouldn't Float*. Captain Rollie has never been easy to get along with, but when we check in aboard his vessel he doesn't even pretend to be happy to see us. He won't look us in the eye and seems distracted. When we ask him if we can start work the next day at six a.m., he says he won't be ready till at least eight, plus a front is predicted. He gives us this news with his first smile.

Day 1. Our scientific contact, Dr. Kathleen Sullivan Sealey, a Scripps PhD on leave from the University of Miami to run the science department at the College of the Bahamas, has promised our rotenone will be delivered from customs sometime early today. We show up shipside at six, Loren's theory being that our promptness would shame Rollie into being on time himself. He is wrong. But today it doesn't matter. The front has indeed come through, with rain and howling winds, and the rotenone never arrives. Boom boxes blare from the conch salad stands and the fruit and fish stalls in the little open-air market under the Paradise Island bridge, and hustlers wander by hawking dope and girls. Finally, the sun goes down.

Day 2. Weather is better, but the rotenone is still impounded and Kathleen dithers about picking it up. I announce I'll go to customs alone if I have to, and in the end we go together. Turns out there is a big, unexpected import duty to be paid. Finally, we haul the big drum of liquid from the customs shed, truck it to the *Guanahani* and are ready to make our first collection in the early afternoon. Mixed with seawater and Joy soap, the stuff works well and we net a good haul from the nearby shallow reef, but in the shower that evening my hair stinks of the oil-based liquid. I never get the smell out.

Day 3. Rollie's mood is worse than ever. William, his deckhand, says it's because he strained his back pulling up the anchor on day two (the windlass was kaput). He's two hours late getting to the boat, which on top of a three-hour trip around the island to Clifton Point means our collection there doesn't start until early afternoon. It's meager: the sixty-foot-deep reef close to the drop-off swarms with voracious yellowtail snappers baited up for the tourist trade. Many of our specimens end up as fish food. Chugging into the closest harbor at nightfall, I have to steer because Rollie's sight fails in the dimness.

Day 4. Miracle of miracles, he's in a wonderful mood when we show up. He's actually there ahead of us for the first time ever. With a 500-kilowatt smile, he announces the *Guanahani's* hydraulic steering system is on the blink. Repairs? Yes indeed. They'll start work very soon. Meanwhile, we can use Fisheries' eighteen-foot Boston Whaler.

We do manage to make one mid-depth site collection in spite of a fierce current that carries away some of our rotenone before it can do its full work.

Day 5. No one is surprised that the *Guanahani* is still hors de combat. The little skiff now contains mountains of our equipment (diving and collecting), Loren, Heidi, me, Sherry (Kathleen's beautiful and very impressive grad student from the University of Miami), Alejandro (a College of the Bahamas intern), and Cass, the boatman. It's got only a few inches of freeboard and would swamp in a good chop. Miraculously, we make two excellent shallow collections and dock back at our hotel after dark. The *Guanahani* remains where she lies . . . For all I know she's still there.

Days 6 and 7. We go flat-out. Up before dawn, Egg McMuffin and coffee at McDonald's, wrangle eighteen tanks and gear into van, shuttle van along coast, transfer to skiff, dive four tanks apiece in up to sixty feet of water, collect on one site, survey another, take DNA samples on foredeck of skiff, back to the hotel after sundown. Preserve specimens. Write up notes. Fall into bed before dinner (sometimes).

Seven collections down, three to go. What would my father have made of it? Collections in the old days always seemed relaxed, unhurried, gentlemanly . . . a bit like an English fox hunt. Jim Böhlke used to complain that the day ended at teatime!

Our frantic pace is not without its psychic costs. "This is one of those L trips," Heidi says, as we gulp our coffee in predawn darkness outside McDonald's. "Lost. Loony. Losing it." We're covered with coral cuts and are beginning to see fish that aren't there. Cass, the skiff's boatman, is even more erratic than Rollie was. He weaves, changes speed, misses, crashes, and doesn't know the water. The only time he's functional is when he's flirting with Sherry, who on top of everything else has a lilting Antiguan accent.

Day 8. For me, the whole project is keyed on many levels to the deep head at Green Cay . . . spiritual being the most important. Our collection there five months earlier had failed, due to the faulty rotenone mixture, so we head out again in our tiny skiff with the weather clear, wind light out of the east-southeast: Loren, Heidi, me, interns Trevor and Valentino from the College of the Bahamas, and Nichola, another University of Miami graduate student to replace Sherry, who's heading home (we noticed her having a farewell dinner with Cass at the Poop Deck the night before). Plenty of manpower. Water and air temperature are both eighty degrees, so going from one to the other makes no difference at all.

We've taken to mixing the rotenone on the dock before we leave, rather than in the cramped skiff. Loren's computer tells him that slack tide falls at ten a.m., so we anchor near the head and wait until then, arguing about numbers and methods and feeling as nervous as if this site were the be-all and end-all of everything. And maybe it is.

The water is fairly clear today. The sun is shining. At ten a.m. on the button, Loren, Heidi, and I go over the side with our bags of rotenone mix and sink slowly through fifty feet to the bottom. *This place reaffirms my faith.* Thank you, Loren. It always does. The first barracuda I've seen in all our recent diving hangs in mid-water about thirty feet away. It's not an old growler, but fairly respectably sized, maybe 3.5 feet. I raise my hand in a salute, but instead of following me it fades away into the blue. Was it ever there?

A cave opens in front of me and I swim inside to let loose the first gout of rotenone from my sack. The cave looks very familiar. One of the first creatures affected is the largest octopus I've ever seen, rambling drunkenly over the sand flats near the head.

Wheels within wheels. Decompressing back on the skiff for another hour's dive, I'm actually not surprised when Loren surfaces, rips out his mouthpiece, and shouts: "You'll never guess what I got!"

There are underwater flashlights these days. At the far end of one cave, in what would have been total darkness fifty years ago, Loren's flashlight beam picked out the second saddle squirrelfish ever to be collected in the Bahamas. The cave narrowed down to the point where he almost got stuck trying to net it.

Day 9. Good shallow and mid-depth collections at Green Cay. Loren and I have had our authority differences on this trip, but by now I'm meekly taking orders and liking it. He's done a great job under almost impossible circumstances. Best of all, the saddle squirrelfish very dramatically suggests that species diversity in my father's resting place is still intact. My notes read: "We did it all! I'm completely wiped out, but kept up w. Heidi and Loren, 20 years younger." In fact, we're acting like kids in a playground.

PART 3

The Chaplin Project

Fish Out of Water

December 2008

LOREN PAINSTAKINGLY SHEPHERDED THE SPECIMENS we collected back to the academy, and they began to gather dust on the shelves beside my father's and Jim Böhlke's specimens. It would take months of work to get them classified and sorted to the point where they could be used in any rigorous comparative study. Loren promised to apply for a grant to pay for this time, but first he had to complete his thesis.

He estimated this would take two months. I should have remembered that he'd already put in ten years, going back long before I met him.

Two more years passed. The archival films touted by John Lundberg proved impossible to use without costly restoration and digitizing. His thesis still unfinished, Loren moved to Cape Cod, had another child, and got a job with a Narragansett Bay marine conservation group. He filed no grant applications. And voilà! My second project leader followed Dominique out of the picture.

Meanwhile, big changes were going on upstairs. President James Baker, the bureaucratic hatchet man from NOAA who had fired Dominique and several other curators and had brought academy morale to an all-time low, was fired himself.

Just before Christmas, I sit down to lunch in the academy president's office overlooking Logan Circle. Out the tall Victorian windows, past the Swann Memorial Fountain, across the Benjamin Franklin Parkway, I can see the imposing Greek Revival architecture of a couple of Philadelphia's more hoary and treasured institutions: the Free Library of Philadelphia and the Philadelphia Museum of Art, where my parents attended balls and schmoozed with dignitaries.

We sit at a round table surrounded by shelves, cabinets, and tables of academy memorabilia, including the skull of a bison shot by Buffalo Bill Cody, a tray of scallop shells collected in the Antarctic by Ernest Shackleton, and a jar containing a specimen of a minnow collected by Charles Lucien Bonaparte, Napoleon's nephew.

Across from me is the academy's new president, Bill Brown, dressed in jeans, a flannel shirt, and a well-worn fleece for warmth. He's about my age, lean, and intense. Lunch is not usually part of his schedule. He works out instead. I'm uncomfortably wearing a suit.

John Lundberg, also in jeans and a fleece, is eating with us. We talk about my father's friend and colleague Jack Randall, the same who jettisoned his gear to escape from a cave after collecting a rare fish. Before coming to the academy, Bill Brown ran the Bishop Museum in Honolulu, where Randall works; he arranged for Jack to continue there long after retirement age. He also reversed a series of budget deficits to get the museum back on its feet.

Why would he leave a place like the Bishop to come to dreary, stodgy Philadelphia? He grins sassily: "I like a challenge."

John Lundberg grins too. I've never seen him look so happy. He loves Bill Brown.

This meeting would never have happened before Bill Brown showed up, but, wonder of wonders, he has encouraged it. I'm proposing an academy fellowship, funded by money I inherited from my mother. The fellow would continue where Loren left off, finally writing the fifty-year retrospective paper comparing our Nassau collections to the original Böhlke-Chaplin ones, and broadening the research that Dominique initiated four years earlier into other areas—different habitats, for example. And our population estimates could fit into the new Global Marine Species Assessment program that determines the danger of extinction for each species.

"I think this might have legs." Brown looks at Lundberg, and Lundberg nods. "What kind of term are you thinking about?"

"I don't know. Four years?"

"Make it five."

We move quickly to the details. The academy would provide office space and lab facilities and raise funds for more expeditions. John Lundberg would advertise the position, run a search, and oversee the fellow for his or her term. Beside his scientific credentials, Bill Brown holds a degree from Harvard Law School. He claps his hands. "I'll have the papers ready in a few days."

I can hardly believe it's true. After four years of uncertainty, the project is actually going to get done. Even better, it's going to be broadened. Bill Brown is my hero! I'm finally on the academy radar screen, and, surprise! . . . it feels like coming home.

Outside, on the cold December sidewalk, I call Sarah in New York to tell her. "Oh, baby, that's wonderful! Your father would be so proud! Come home and we'll celebrate." I can hear our two-year-old daughter, Rosie, gurgling in the background.

"I've just got to drive out to Haverford and tell Nancy. I'll get on the turnpike from there."

Most of my parents' Philadelphia friends are either dead or in nursing homes. I've kept in close touch with my favorite one, Nancy Grace, an eccentric writer and traveler who'd been my mother's bosom buddy and reminded me of Auntie Mame—always up for an adventure, especially if it involved a good party. Alone of all my mother's crowd, she seemed to appreciate, even enjoy me. I saw her as a kind of proxy for my mother, providing the approval and support I never got from the actual thing. If anyone should hear my news, it's Nancy.

She'd visited me in Saigon when I was a *Newsweek* correspondent there in the late sixties. Some of the dialogue from then is etched into my brain forever. Driving in from the chaotic, war-zone airport in the bureau's Toyota, bumping past sandbagged checkpoints, and swerving to avoid APCs (armored personnel carriers), she observed: "This is all terribly interesting, you know."

"Yes. Are you planning to stay long?"

"Long enough to see for myself what's going on. Some of you journalists are saying we're losing. What do you think?"

"I'm afraid we are."

"Well, I mean, how does one tell? Can I get into the field?"

"I don't think so, Nancy. You have to be accredited. Maybe I can get some people together for you to talk to."

As far as battleground viewing went, the best I could do was to drive her to Cholon to rubberneck around the rubble and bullet holes left from Tet. "This doesn't look nearly as bad as I was led to believe," she sniffed, inspecting the place where the famous photograph was taken of General Nguyen Ngoc Loan executing a captured Vietcong with a pistol to the head. "Not *nearly*."

In big, white-rimmed dark glasses, a diaphanous lavender shift, and high heels, almost six feet tall, she cut quite a figure during cocktail hour on the terrace of the Hôtel Continental, and after that it was easy enough to arrange a dinner party in her honor. My colleagues—young journalists, junior diplomats, and spooks a year or two out of the Ivy League—all loved her, and by association my own stock in trade went up considerably. We were invited to many more parties, and after a few days she flew happily back to Singapore. I never found out what she'd decided about winning or losing.

Nancy's nursing home is at the end of the same road we used to live on, College Avenue, in the Main Line town of Haverford.

My mother's family seat, a big white stuccoed fieldstone house with green shutters, still broods on its little rise up the treelined drive, but the rest of the twelve-acre property has been subdivided into tiny lots, each with its own McMansion. The sight doesn't affect me one way or the other. This place has never been my home.

Nancy, the world traveler, is now ninety-seven and has been bedridden for years. A stuffed green parrot in a gilded cage sits on the dresser: Leticia, the heroine of her best book. A manuscript she's been editing lies on the bed beside her. She's asleep.

"Did you call ahead?" the nurse says. "She likes to be prepared, get ready, look her best."

I shake my head. "This was kind of spur of the moment."

"Well, it'll be dinnertime in half an hour. I'll wake her up." Touching an emaciated shoulder. "Nancy? Nancy, you have a visitor."

I watch my last living link to my parents' generation slowly surface as if from a deep dive. Her papery eyelids flutter, close, flutter, close. When they finally open, her large gray eyes are looking right at me. A smile breaks over her face, and she reaches out her arms like a child. "Oh, *Charlie* . . ."

The truth suddenly dawns on me. I've been a stand-in for my father from the beginning. She's had eyes for him all along. Was this the real reason she was so friendly with my mother? I remember that after the fatal car crash, Nancy had glommed on to any excuse at all to show up at our door. Unfortunately for her, my father had other ideas: three months after the crash he married another family friend, Agnes Osborne Trimingham. I would have much preferred Nancy to the uptight, controlling Agnes, but maybe he saw her as too much of a handful.

I take a hand and squeeze it a bit coolly. "It's Gordon, Nancy."

The smile fades a little but doesn't disappear. "My dear. You're looking *so* much like your father. He was so handsome, wasn't he?"

"He was indeed. But you're looking good yourself."

She puts her free hand to her hair. "Oh no. No, I look *terrible*. I wish you'd called so I—"

"I've been at the Academy of Natural Sciences, Nancy. I have some good news to tell you."

Her reaction is not exactly the one I expected: "Well, what are you going to call it?" she asks in a businesslike way.

"Call what, Nancy?"

"The *fellowship*. You get to name it, don't you?"

"I guess I do." That topic hadn't come up at the lunch. I'd never even thought about it.

"I like 'the Chaplin Fellowship,' don't you? Simple and right to the point. It's your mother's money, your father's work, and your . . . ah . . . well, without you it wouldn't exist, would it?"

"My idea?"

"Your *doggedness*. You've turned *dogged*, my dear. You are now a *dogged* person. I do love that word, you know. And I don't even particularly like dogs. Isn't that strange? I like parrots and elephants. Did I tell you I'm working on a new book? There's a piece about Afghanistan in the early sixties before all the trouble. God, it was heaven then. I hope Obama gets to read it." We're still holding hands, and she gives me a squeeze. "Oh, Gordon. Do you remember Saigon?"

"Nancy, I'll never forget it. You were the belle of the ball. They just couldn't believe you."

"Oh, come on, now. Really?"

"Absolutely." We've been over the story many times, and at this point she always looks as proud and pleased as a debutante on the morning after her coming-out, hearing about her first whirl around the dance floor in her father's arms.

"Well. It was one of the best dinner parties I've ever been to. All those parties were. My God, those young people were all so attractive and charming, I almost forgot what I was there for."

"Did you ever find out? You never told me."

This is new ground. A sly grin. A sideways glance. "Well, of *course* we were losing. I knew that before I came . . . I'm a Quaker, for God's sake. But I wanted to see you in action and report back. Your father talked about you constantly, you getting the word out, a very,

very important job. Knowing him, he probably didn't tell you, but he was awfully proud of you, you know."

I feel the same expression of pleasure and pride settle over my face as I saw on hers a few minutes earlier. She saved this up to tell me now, for some reason, though I know she's planning to live to a hundred and she has plenty of time left.

Wrong, as it turns out. She died at ninety-nine.

Dr. Ilves

October 2009

THE NEW CHAPLIN FELLOW, DR. Katriina Ilves, thirty, is on familiar terms with all the dizzying complexities of modern taxonomy. Her degree in fish systematics and biogeography ("study of the historical processes that may be responsible for the contemporary geographic distribution of individuals") is from the zoology department of the University of British Columbia, one of the world's best. Her thesis used molecular applications to analyze the evolution of smelts across the northern oceans.

Her doctoral advisor called the thesis "a classic accomplishment in Holarctic systematics and comparative biogeography" and noted significantly that this kind of work could lead to a better understanding of how recent global changes impact species diversity. Katriina herself wrote the academy search committee that she hoped to start using new species distribution models to gauge the impact of future climate change for conservation planning and marine-protected-reserve design in the Bahamas.

This ecological dimension was the main reason for my enthusiasm with Katriina's candidacy.

Traditional "hard science" fields like ichthyology, botany, ornithology, et cetera, have always concentrated on classification, but since my father's day simple taxonomy has evolved toward systematics, the study of how each living organism over time fits into the tree of life. In the sixties, computers and DNA analyses enabled the evolutionary approach to supersede simpler classification based on shared characteristics, and many scientists have become obsessed with the way the traditional building blocks can be reorganized. A lot of prestige and grant money is at stake.

In all the trips and meetings that led up to this book, I've had many conversations with scientists about the relative merits of the science of ecology versus the science of systematics. It's Aldo Leopold versus Charles Darwin, to coin a phrase. Let's face it, Darwin's science is more significant in the long run, more sophisticated, more challenging. *The Origin of Species*, after all, changed our view of the world and created an argument that's still passionately being pursued. All of the scientists I worked with were systematicists first and ecologists second—more prestige in "hard science." Dr. Mark Westneat, curator of fishes at the Field Museum in Chicago, who was on a later trip, put it this way: "Look, I agree that the planet needs to be saved . . . the reefs, the rain forests, the polar ice fields, and all the rest. But that doesn't take science, it's just common sense. It's not intellectually challenging. In systematics, you have the potential to discover something completely new."

But our study is an ecological one. Evolution, classification, and heredity are not its major concerns. We're not dealing in thousands of years but simply with what happened in the last fifty, thanks to the baleful influences of just one of Darwin's species: *Homo sapiens*.

I personally care more about preserving a species than I do about the way it's classified, though I grant that the two endeavors might depend on each other in the end. I favor field observation over molecular analyses.

My hero, Aldo Leopold, who lived from 1887 to 1948, was instrumental in the development of ecology as a new science, though, as one biographer puts it, "it is probably safe to say that he was attracted more by his love of the outdoors and the excitement of the conservation movement than by the intricacies of ecology." He cut his teeth as a US Forest Service supervisor in northern New Mexico, responsible for the million-acre Carson National Forest. Originally, he was in charge of exterminating wolves to save the deer. Then he realized that wolves were a crucial factor in the overall equation. Hence his famous essay, "Thinking Like a Mountain."

In the end, to restore threatened habitats and preserve some kind of ecological integrity in the landscape, he was aiming for nothing less than "an internal change in our intellectual emphasis, loyalties, affections, and convictions": a new land ethic. Which could just as easily be applied to the oceans.

Katriina is a dual-passport Canadian-American, of Estonian descent on both sides. She has black hair, usually dresses in black, and her skin is the palest I've ever seen.

"What do you do about sunburn?" was one of my questions to her in our selection committee meeting. I was thinking of Jim Böhlke's flaming sunburns and tattered skin.

"Use a lot of sunblock," was her good-humored answer.

The fact that she has little knowledge of Caribbean fishes and not much field experience does not seem to be a problem either. "It won't take me long to pick that up," she says. She's very at home in a lab; that's the most important thing in today's scientific world.

The truth is that I find Katriina a bit daunting. She speaks a language that's almost impossible for a layperson to understand. In this, she was no different than the other candidates we interviewed . . . just more highly qualified.

The Study Finally

May 2010

LIKE THE WHEELS OF JUSTICE, the wheels of science grind slowly. My father's tome *Fishes of the Bahamas* began as a two-year project for a small handbook on new species. Fifteen years later it became a 771-page definitive work.

Katriina finishes her paper in May 2010, almost exactly six years after the first of our five expeditions to the Bahamas, and it is accepted after minor revisions by the prestigious *Bulletin of Marine Science.* As of its publication in July 2011, the Chaplin Project is part of the official record.

Of course, there have been other studies made of the changes in reef fish communities over the years, but none of them cover a fifty-year span (most cover less than thirty years), and none involve the use of rotenone for sampling. They all rely on visual surveys. So my father's and Jim Böhlke's collections form a unique baseline, as Dominique and John Lundberg correctly foresaw.

Dr. Jeremy B. C. Jackson, the éminence grise of baselines and their limitations, weighs in thusly: "Conventional ecological data are clearly inadequate to measure the ecological impacts of fishing or any other long-term human disturbance. Most observational records are much too short, too poorly replicated, and too uncontrolled to encompass even a single cycle of natural environmental variation."[1]

Thanks to the Böhlke-Chaplin collections, we've been able to counter this argument. We have some replication, good controls, and the longest time span available for a rigorous comparison.

Since the study's aim is to document changes in reef fish communities over fifty years, the big underlying issue—the gorilla in the elevator—involves what's come to be known as the philosophy of biodiversity. Aldo Leopold was a pioneer in developing this new concept: wolves are as necessary for the health of the environment as deer are. Since the natural order is imperiled by man, man has to be the custodian. He should not seek to exterminate wolves but to preserve them . . . even though they eat his cattle and don't appear to help his cause. The natural order of things is best.

Biodiversity historians maintain that man has been threatening the natural order of things ever since he appeared on the scene in the Stone Age, about two million years ago. About 10,000 years ago, at the end of the Ice Age and the beginning of the modern Holocene era, the planet entered the Holocene Age of Extinction, which is still going on. In fact, certain radical factions of the biodiversity community predict that, at the present rate of extinction, most species on the planet will be gone in one hundred years. Whether this is good or bad depends on such things as your political affiliation. In other words, it's a moral issue, not a scientific one. But it is a fact that man, alone of all the planet's creatures, has caused quite a few other species to disappear for good.

I, for one, side with my hero, Aldo Leopold. Anyone capable of writing "Thinking Like a Mountain" has to be in touch with the way the world really works. Another famous quote from him: "Everybody knows that the autumn landscape in the north woods is the land, plus a red maple, plus a ruffed grouse. In terms of conventional physics, the grouse represents only a millionth of either the mass or the energy of an acre. Yet subtract the grouse and the whole thing is dead." Which brings us back to the question of species diversity on Bahamian coral reefs.

Let Katriina introduce the work (I've edited her prose for this context):

> Caribbean coral reefs have been in decline for decades. The causes of these declines are both natural and anthropogenic, and include hurricanes, coral disease, mass mortality of invertebrate algal grazers, pollution, physical habitat destruction, nutrient runoff, and warming water temperatures. A question of particular interest in coral reef community ecology is how reef fish communities are affected by such widespread coral habitat degradation. . . .
>
> A focus simply on diversity measures may miss important changes in community composition, such as the loss of dominance of particular trophic groups. Furthermore . . . a time lag between coral death and the subsequent effect on the surrounding reef fish community may obscure the true impact of disturbance, such as bleaching, in studies with data collected immediately following such an event. Studies examining the effect of coral loss must therefore take into consideration the timing and duration of habitat disturbance events when interpreting the results of fish community structure analysis.
>
> Although quantitative assessments of coral cover around New Providence Island pre-1970 are lacking, examinations of photographs yielded estimates of 30%–40% in many areas of the Bahamas, and a coral survey contemporaneous with our study estimated coral cover at 2%–20% . . . suggesting declines in coral cover

> have occurred over the time frame encompassed in our study. . . . The historical Böhlke and Chaplin collections, therefore, provide an important set of baseline reef fish community structure and distribution data that is quite rare in marine systems.[2]

Accurate identification is the basic foundation of any study, so in her first six months on the job (October 2009 to March 2010) Katriina wrestled more than 10,000 individual specimens into workable shape. In their fifteen years of collecting from the four selected sites, Jim Böhlke and my father netted 5,348 individuals for the academy archives. Amazingly enough, on our two collecting trips in 2006, we netted 5,423.

Partly due to the new system of phylogenetic systematics, many of the species collected by Böhlke-Chaplin fifty years ago have acquired different names and sometimes different evolutionary histories. Katriina updated them as much as possible, and in uncertain cases she went with the historical names.

She excluded free-swimming fish, which don't live in the reef, and very rare fish collected only once at one location. She excluded our many unsuccessful collections, made with poorly buffered rotenone, and historical collections that were noted as "poor kills" or otherwise compromised. She excluded all questionably identified specimens. With all these exclusions, she ended up with 4,614 historical individuals and 4,857 recent ones.

I held my breath at her next step: if it didn't work out, the whole study would fall apart. This was to compute "species accumulation curves" that graph the number of new species against the number of individuals collected. As the number of individuals increases, the number of new species should level off. If it doesn't, the data is incomplete and can't be used without further collections, impossible for the historical stuff.

Miraculously, for our four collecting sites, historical and current data followed almost exactly the same curve, leveling off nicely at about 150 species for the full count of 9,472 individuals. When I heard this news, I bought yet another bottle of "the widow" (as my English grandfather used to call Veuve Clicquot champagne) and drank it with Sarah.

The study includes a grim assessment of the fishes' habitat, coral reefs. Only local fishermen seem unaware that Bahamian reefs are in as bad if not worse trouble than reefs in other parts of the world, especially the Caribbean. "Doomed" was the verdict of one of our expedition coral experts who studies reefs off Fort Lauderdale.

Global warming, disease, bleaching, pollution, rampant algae, hurricanes, overfishing, and overdiving: the list of causes goes on and on, and in one way or another mankind may be responsible for all of them. A recent report from the Global Coral Reef Monitoring Network lists global warming as the top threat and estimates that the El Niño condition of the mid-nineties destroyed 16 percent of the world's reefs within two years.

El Niño conditions happen when trade winds in the Western Pacific lose their force, allowing surface water to warm up. The warm water drifts east toward South America, blocking normal cold upwellings and causing water temperatures to rise worldwide. The condition happens every four to seven years, but the 1997 to 1998 El Niño was by far the strongest of modern times. Some scientists claim a connection to global warming and predict more and stronger El Niños as the climate continues to change.

A rise of more than one degree centigrade in your body temperature will give you a fever. In the ocean, it can kill coral.

Warming water puts sensitive coral under stress and affects its ability to nourish the tiny algae (called zooxanthellae) that live inside its polyps and in turn nourish the coral itself through photosynthesis, like leaves on a plant. The stressed coral expels its algae (or they expel themselves), turns snowy white, and is much more susceptible to disease. Some corals can eventually get their algae back, but most end up dead.

Walter Jaap, on our July 2006 trip, concluded that the highest percentage of live coral cover on any of our sites was only 20 percent, and this was at the shallow site at Green Cay. The average among the other eight sites was 5 percent. This is in sharp contrast to the condition of reefs before 1970; he estimated that in those days the average live coral cover in the Bahamas was greater than 30 to 40 percent.[3]

Jaap cited a study of aerial data from 1943 showing 214 patch reefs in one area near Nassau Harbor; the number had dropped to 133 by 2004. And he cited a more general 2005 study listing Bahamian reefs as over 60 percent degraded, behind only Jamaica and Panama. Jaap attributed this sorry state almost entirely to the tourism industry, "which provides an estimated 60 percent of the GDP and employs 50 percent of the workforce, making it a powerful economic force, often at the expense of the surrounding ecosystem."

Jaap's paper strangely didn't include two other major factors in the death of the reefs: a 1983 epidemic that killed off the algae-eating black sea urchin (*Diadema antillarum*) and thus drastically altered the precarious balance between algae and live coral, and, most importantly the El Niño condition of the mid-1990s that raised seawater temperature more than a degree centigrade above average in the Bahamas to cause widespread coral bleaching.

A study of remote and pristine reefs near the Bahamian Out Island of San Salvador (far from the tourist trade) conducted before, during, and after that El Niño, describes "the rapid transition of this reef community from one in which corals and algae were co-dominant to a community dominated by macroalgae." The authors note that this change occurred with negligible input from the local population. The incredible thing is that it occurred in only four years: 1994 to 1998.[4]

Katriina's basic conclusion, after analyzing species accumulation curves and running complicated multivariate analyses comparing sites, depths, and time periods, is at first glance wonderful news. Even with severe coral degradation, *she found no measurable difference in species richness over fifty years!* The original Böhlke-Chaplin collections included 4,614 individuals representing 152 species, and ours had 4,857 individuals representing 148 species. The differences are too tiny to be reliably counted.

My own eyes had told me this during our collections—and there was the saddle squirrelfish—but I hadn't dared to believe it. The study confirms this, though. In a grim picture, there is hope. Other studies have described a generally declining fish population, but as long as the species themselves survive, the means for regeneration are still intact. Extinction hasn't happened yet. Biodiversity is preserved, at least for the time being.

But the time qualifier here is very important. Even though diversity is pretty much the same, Katriina found crucial changes in the proportions of the community: fewer plankton-eaters like the tiny red cardinalfish and more algae-eaters like parrotfish and wrasse:

> At face value, the results of our study are consistent with observations that the structure of coral reef communities has changed relatively little despite increases in coral mortality and algal cover, overall declines in fish abundance, overfishing of large predators, pollution, and loss of nursery habitat. This interpretation, however, is an oversimplification of what is possibly an initial stage in the response of reef fish assemblage to recent habitat disturbance. . . .
>
> A lack of change in species diversity when reef structure is maintained has been explained by observations that even dead coral continues to provide protection and food resources to small reef fishes; however . . . significant shifts in community composition can occur after coral bleaching even with no apparent change in diversity metrics. Our observation of the increase in herbivore relative abundance, a signature in multiple cases of a response to a relatively recent (<10 yrs) loss of coral, provides a clue regarding the

> potential time frame during which the coral died, algae colonized, and reef fish responded.
>
> Additional data are needed to test whether the coral habitat is no longer functioning well as a shelter from predators. . . . Regardless of the cause of mortality, it is well documented that dead coral will retain its structure for only a limited time, after which it will be eroded.[5]

An initial stage? And what comes next? The implications are terrifying.

The Debt

It so happens that for all our collecting sites, the skeletons of the reefs still rise pretty much intact from the seafloor, even though the coral itself has died. But in many parts of the Caribbean dead reefs are crumbling fast. One comprehensive report on reef architecture, covering five hundred surveys on two hundred reefs between 1969 and 2008, concludes that the most complex structures of forty years ago no longer exist. An initial period of decline occurred in the early eighties. A second began in the late 1990s, was accelerated by El Niño in 1997, and is still going on.[6]

So, finally, we come to an almost unbearably poignant situation. Species richness might have maintained itself against all odds over the last fifty years, but it is living on borrowed time. The study chillingly calls this time period—between the death of the coral and the actual disintegration of the reef—a "degradation debt." Katriina points out that the debt's time frame depends on "numerous factors, including the coral species and duration of disturbance." But in most of the Caribbean, structure had begun to flatten by the 1980s and now the most complex reefs have all but disappeared. Our own study showed a decrease in cardinalfish, which are more dependent on reef structure than other species: the debt may already have started to be paid around New Providence.

There are a few rays of hope. Modern coral reefs started growing 10,000 years ago at the beginning of the Holocene period, and it's easy to think of them in almost eternal terms: now that they're in trouble, another 10,000 years will be required for them to recover.

Actually, recovery could take a lot less. Counterintuitively, reefs are much more sensitive than fish are to destructive and/or benevolent ecological influences. The El Niño study in San Salvador, for example, showed radical change for the worse over a period of only four years.

The reefs off Jamaica, where the ecological study of coral communities began in the 1950s, are a "classic example of reef degradation in the Caribbean," as one study put it. But in the last ten years at least one miracle has happened there, off the north coast near Discovery Bay, on a shallow reef known as Dairy Bull.

Scientists surveyed this reef in 1995, 2003, and 2005. They found that over these ten years live coral cover at Dairy Bull Reef had nearly doubled, while algae cover was reduced by 90 percent. The most dramatic coral cover increase was in staghorn (*Acropora cervicornis*), the species most affected by general degradation. We found no staghorn at all in our trips to the Bahamas. A 2005 photograph at Dairy Bull shows huge thickets of it growing in and around big heads of boulder star coral (*Montastraea annularis*), and the scientists compared the general health of the place to the reefs of the 1970s and earlier.[7]

The big boulder star coral heads probably made the difference, the scientists hypothesized. Possibly because the massive structures are more resistant to hurricanes, they had mostly survived at Dairy Bull over the years, while less massive corals at other locations had died and crumbled away. Among other things, the healthy heads provided the extra structure needed for a large population of algae-eating sea urchins. "While our study is the first to report a phase-shift reversal on a Caribbean reef," the author writes, "our results are spatially restricted to the reef at Dairy Bull. However, when interpreted in conjunction with other studies [which show how sea urchin population goes hand in hand with reef health], they appear to provide some important clues about [what] factors [are] responsible for promoting coral reef recovery throughout the Caribbean."

Either the cup is half full or half empty: Dairy Bull's miraculous recovery could be an anomaly, or it could show the way to better days—the scientists can't say for sure; all they can report is what they observed at this one location. Why did boulder star coral flourish at Dairy Bull and not at other reefs? Why did the general coral recovery on Dairy Bull just happen to start in 1995? These imponderables could cut either way.

Hope Springs Eternal (2)

THERE ARE NOT SO MANY imponderables in a 2006 study published in *Science* magazine by a group of talented young scientists working at the Exuma Cays Land & Sea Park about forty miles southeast of Nassau. "Few reserves are either large, old, or effective enough to have had a significant impact on large predators," the authors write. "An exception is the Exuma Cays Park . . . one of the few places in the Caribbean where the long-term impacts of reserves can be investigated."[8]

A marine protected area (MPA), as these reserves are technically called, is simply a part of the ocean set aside for noncommercial use, like a terrestrial park. Many of these parks have been set up by governments, of course, but a reserve is only as effective as its enforcement system. It's not uncommon to find that this consists of one man in one outboard launch responsible for hundreds of square miles.

The Exuma Cays reserve is the second oldest marine park in the world, after the Fort Jefferson National Monument in the Tortugas, just west of the Florida Keys. It was set up in 1959 by my father, Ilya Tolstoy, Jack Randall, Andy McKinney, and a few friends. Now there are many others, and most authorities agree that the reserve system is the last best hope for the silent world.

My father's diving buddy Count Ilya was the driving force, and a pretty unlikely one. The grandson of Leo, he'd emigrated to the United States in 1924 at the age of twenty-one, not long after the Bolshevik revolution. By 1931, he'd found his way to Nassau, and except for a wartime stint in the OSS, during which he led an expedition from India across the Himalayas to Tibet to enlist support for the Allies from the Dalai Lama, he was a Bahamian regular until his death in 1970.

He dabbled in filmmaking (at the age of ten, I was a costar in one of them), founded Marineland in Florida, invented a hypodermic harpoon for the live capture of large sharks, perfected designs for underwater cameras, and took up with a fabulously rich Canadian widow whom he couldn't marry under the terms of her husband's will without forfeiting her inheritance. I remember his pelted barrel chest, tree-trunk legs, and mysterious ability to sink through the water at will without weights.

Returning to Nassau after the war (about when we bought the Chaplin House), Tolstoy took a trip to Andros and was "shocked by the lack of underwater life. . . . [I]n many areas where large conchs were abundant, now one sees only a few small ones here and there. . . . On land, the mass morning and evening flights of white-crowned pigeons were thinned out in ranks by open season and no bag limits. I saw the ghosts of Passenger Pigeons in the air. There were fewer iguanas, and they were not as tame. I saw some dead ones lying around with .22 bullet holes in them."

Tolstoy's baseline of experience in the Bahamas went back twenty years before my father's. That might explain my old man's ambivalence when he wrote in his diary in 1957: "Had a long talk with Ilya, who is very keen on forming a preserve in the Exumas . . . Warderick Wells or thereabouts. I don't quite understand how it will work and it seems a bit far-fetched. However, one must look ahead these days, very far."

So they joined forces, and in the next few years secured support from the American Museum of Natural History in New York, the Nature Conservancy in the UK, and the Bahamas Chamber of Commerce. The "Bay Street Boys" passed an act to set up an administrative body called the Bahamas National Trust. Finally, 176 square miles of reefs on the leeward side of the Exumas were leased to the trust.

It took another twenty-seven years, long after Tolstoy's death and Bahamian independence, for a ban on commercial fishing to be enforced with several wardens at the reserve, but the results have been incredible. Among other things, the 2006 study reported that Nassau groupers, a favorite of the local fishermen, were seven times more numerous in the park than outside it.

Groupers are voracious fish-eaters, and one of the study's aims was to find out if the reserve's burgeoning grouper population was affecting the population of a crucial algae-eater, the parrotfish. Since the 1983 sea urchin plague, parrotfish have assumed most responsibility for keeping the coral-killing algae in check.

Parrotfish are also targeted by local fishermen. Would their population have grown inside the protected reserve? Or would it have been held in check by the fish-eating groupers?

The study concluded that most groupers aren't big enough to eat parrotfish. Inside the reserve, the parrotfish population was growing: it was double that found in the neighboring reefs and five times that of other locations in the Bahamas.

The scientists then compared the condition of coral inside the preserve with its condition in other places where parrotfish were less numerous. "Our data reveal a strong negative relation between fish grazing intensity and macroalgal cover in the Exuma Cays; the cover of macroalgae was reduced fourfold inside the reserve, whereas there were no reef-scale fluctuations in cover in systems with no reserve."

Their study is the first documentation of the "small-scale" effects of a reserve. The populations of heavily fished species like the grouper and the parrotfish go up. Parrotfish grazing goes up and reduces the algae cover. Four times less algae means four times more coral cover. More coral cover means better reef structure. Better reef structure means increased fish populations and intact species richness. And on and on: it's called a "trophic cascade," (*trophic* meaning the level an organism occupies on the food chain). Outside the reserve, in the absence of parrotfish, algae reign supreme. The relationship between algae and coral is a zero-sum equation.

A high-powered posse of senior scientists, including the famous critic of shifting study baselines Jeremy B. C. Jackson, recently hailed no-take reserves as "the best management tool for conserving coral reefs and many other marine systems." Reserves can't reverse global warming, but they do foster healthy coral that can better withstand the effects of bleaching. Their fecundity produces not only bigger populations of adults but also more coral and fish larvae that can migrate to other areas. Jackson et al advocate chains of reserves close enough together to protect even the most stay-at-home species.[9]

The Bahamas government is rising to the challenge: as I write, it has created seven more reserves covering more than 650 square miles. My father and Count Ilya would be very, very happy.

My own baseline goes back fifty years. Jeremy Jackson argues that, for a realistic picture, you really need to go back to the first available detailed human records, say from the eighteenth century. "Studying grazing and predation on reefs today is like trying to understand the ecology of the Serengeti by studying the termites and locusts while ignoring the elephants and the wildebeest."[10]

Well, maybe so. But our study found that changing fish communities were responding to a *recent* disturbance: this had to be the El Niño of 1997–1998. Stuart Cove's video transect of the reefs near Goulding Cay off the west end of New Providence Island, taken just before the disaster, backs this up, as does the San Salvador Island study. A 1998 survey of the Andros Island barrier reef has this to say about it:

> Local observations of "white coral" were first reported in early July (1998). One month later we found many massive colonies with either partial or complete recent tissue

> mortality and signs of disease. . . . At severely impacted fore reefs, we estimated that up to 50% of the live stony coral cover had recently been lost. We believe severe outbreaks of one or more diseases must have occurred in late June/early July and moved through the entire area at a very rapid pace, perhaps in response to the increased seawater temperatures. The unusually luxuriant growth of the *M. annularis* species (boulder star coral) may have made these communities very prone to outbreaks of disease. Recovery from this massive disturbance will take many years, possibly decades.[11]

El Niños will be happening more and more often, according to the Intergovernmental Panel on Climate Change, the leading international body assessing global warming. Unfortunately, reports from its thousands of scientists have mostly been ignored in the US. If and when another El Niño occurs on the scale of 1997–98, it will be back to ground zero for whatever recovery the reefs might have made. And maybe even farther back.

Full Fathom Five

Rest in Peace

November 2009

SARAH AND I BOARD A chauffeur-driven golf cart inside the Lyford Cay gates. It's a truly magical Bahamian evening: three-quarters moon rising into a velvety sky, cool, moist southeast breeze rustling the palms. A Nassau truism: the weather's always good for the BREEF Ball.

Cabin lights of the docked yachts at the marina glimmer on the still water as our golf cart ferries us along the embankment and down a long stately drive lit with flaming torches. We disembark near a man in a beautifully disheveled linen suit, who hands us a serving of rum punch in a coconut shell and bids us welcome.

My father would have loved the fairly daunting coincidence that the theme of this evening, "Return of the Drumbeat," is exactly in tune with what I've been trying to do since my first expedition here with Dominique. The BREEF Ball's baseline comes from fifty years ago too, its aim "to recapture the lively innocent spirit of that era—both to rekindle memories and to give a glimpse to those who missed them . . . and by extension, to remind us of the richness of the Bahamian sea of those storied days when the abundance of the reefs seemed eternal."

In fact, it's my father's evening, an eerily perfect fit. The elderly entertainers—Peanuts Taylor, Ronnie Butler, Chippie Chipman, Freddie Munnings—would have played for him at the Bahamian Club, the Drumbeat, the Junkanoo. Some of them have even played for me.

The revelers in the grassy, torch-lit, open-air cocktail area dress and act like my father's friends did fifty years ago, though they're a different generation . . . mine, even my older daughters', who are now pushing forty. Many of the faces seem familiar. My childhood buddy Freddie Wanklyn would be here if he hadn't moved to France. I bet Heather Bethell is around. And Bobo Sigrist? Could that be my childhood love, Carol Salmon, examining the dolphin photo? Or is it Carol Salmon's daughter? Should I go up and introduce myself?

It's a homecoming, in kind of a silent way. I feel quite a bit the way I did on my first returning SCUBA dive on the deep coral head at Green Cay. And, incidentally, the species richness here at the gala looks as if it's been just as well-maintained. In fact, it's been added to: about a third of the faces are black. Michael Braynen himself is probably here, though I don't see him.

Months earlier, Casuarina McKinney (BREEF's executive director since Sir Nick died of lung cancer) asked if I had some old underwater footage from the fifties to show at the ball, on the two six-foot-high video screens at each end of the outdoor cocktail area. It so happened I did: two sixteen-millimeter reels shot mostly by my father, some of the first underwater movies ever made. I had them transferred to Beta stock and copied onto DVDs for donation to BREEF.

One was in fact the short film that Ilya Tolstoy had produced in the early fifties—"Treasure of the Bahamas"—shown in theaters across the country before the main feature. Freddie Wanklyn and I had a brief moment of fame as schoolboys who play hooky to hunt for treasure, but end up spearing a boatload of crawfish and selling them for good money at the Rawson Square market.

Sarah and I glue ourselves to the two big screens, now showing old footage of the gala's famous entertainers performing in the Nassau clubs of the fifties. We're back to back: she's watching one screen, I'm watching the other. At any minute now Freddie and I will be up there, aged twelve and ten, diving through lush stands of elkhorn coral, among clouds of fish, Hawaiian slings cocked, boating one crawfish after the next. Forever young. Forever in paradise.

But nothing happens. Could it be that Casuarina has edited them out? Maybe she never got them? As the minutes drag by, I feel my breath shortening, the palms of my hands beginning to dampen.

"Let's go for a walk," I finally mutter, and as soon as we start moving I feel much, much better. Suddenly, I realize I don't want the partygoers to see me as a kid in paradise long ago. It has to do with Aldo Leopold and preserving bright, shining memories. It also has to do with putting ghosts to rest.

The nearby dining area is out of a fairy tale, too, forty tables of eight dotted about a broad lawn overlooking the marina, lit with large paper globes like Japanese lanterns, the tables covered with real old-fashioned crocus sacking like you used to see at the Rawson Square fish market back in my father's day. Main courses are traditional Bahamian staples, such as coconut rice with black beans, fried plantain, Caribbean chicken curry, pork tenderloin with pineapple, johnnycake, and mango chutney. Dessert is that old Bahamian favorite, guava duff.

My dinner partner is the lively Michelle Symonette (her family includes a prime minister and a sailing champion), who owns a trendy boutique called Fab Finds. She is on the gala ladies' committee and knows all the gossip. If anyone can draw a bead on this gathering, it's her. "So tell me." I give her my best party grin. "What do you think is the . . . ah . . . common thread with this crowd?"

She smiles back indulgently. "Well . . . everyone here has longtime connections." Putting her hand on my arm. "Like you. You grew up here, didn't you? Where was your house? I'm sure I know it."

I tell her. Beside the reefs, my strongest longtime connection here has been the Chaplin House. After the Kosoys bought it, I Google-Earthed it many times to get a satellite image and find out what had happened, but a perverse cloud always obscured the place. "I haven't seen it for a while," I say, afraid to say anything more.

Her hand on my arm tightens then rises to cover her mouth. "My God. It was next to Richard Harris's place, right? To the west?"

I nod.

"It's gone, Gordon. You didn't know?"

"Not really." My plate of curried chicken begins to look like an exhibit in a glass case at a museum. "No."

"It's quite a story actually," she goes on with relish. "Kosoy leveled it, along with every tree and shrub on the property. He was going to build condos, but what he didn't realize was that he'd violated an ordinance prohibiting destruction of any structure over one hundred years old. Which yours was, right?"

"Yes. It was built in the nineteenth century. I think one of the owners was Alexander Agassiz, the naturalist. Anyway, it was called the Agassiz House before we got it."

"Exactly. So, Kosoy had already stepped on a few toes, and his stock in Nassau was kind of low. They slapped him with a violation notice and refused to give him a permit to build his condos. Plus his wife divorced him."

I favor Michelle with a crooked smile.

"You should see the place now." Shaking her head. "Nothing. A completely barren strip in the middle of Paradise Island. A disaster." Staring at me with big, sympathetic eyes. "No, you wouldn't want to see it. You poor man."

But now that I know for sure, the news doesn't knock me down. Just the opposite: I feel a very strange rush. The show band strikes up, and Sarah and I head to the dance floor along with almost everybody else. Bahamians love a good junkanoo, and the band is playing a sizzling mix of old and new with a backbeat you can feel down to your toes.

I look at the moon then look out over Sarah's red hair at the festivities. Casuarina, imposingly pregnant in a black dress and aqua scarf, is standing on the grass beyond the tables, chatting with an older man in a perfectly tailored dark blue suit. His back is to me, but I can see he's about six feet tall, well built, hair dark and wavy. As he talks, he cocks his head and raises a finger in a gesture that's so familiar it makes my heart stop. Then I can see the set of his shoulders, the bend of his elbow, the way his feet move in their slim, long black dress shoes, are all part of the same indelible picture.

The man's my father.

Casuarina's hand goes up absently to arrange her hair. Her smile shows that the man has her complete attention. She cocks her head just the way he did and reaches out to touch his forearm.

I lose the beat. Sarah backs off and looks around to see what's distracting me. All I can do is shake my head and keep staring. I feel dizzy. Am I about to faint in front of everyone?

Then the man talking to Casuarina turns his head so his face is in profile. Of course, he's not my father, how could I have ever thought so? He must be her new husband. Shakily, I gather Sarah into my arms and shed tears of relief.

I'm free as a dolphin.

Where We Stand

The Present

As TO MY FATHER'S LEGACY, the ocean and its inhabitants past and present, this newly minted activist wishes he could do something truly earthshaking to help preserve it . . . such as working to develop a worthy successor to the Kyoto Protocol reducing worldwide greenhouse gas emissions, and forcing the United States to sign this time around. Global warming is by far the biggest threat to the health of the seas: if the skeptics could only see that video we unearthed of the Goulding Cay reefs before El Niño in 1997 and compare it to conditions there now, maybe they'd shut up. But I doubt it. Too much money is at stake. People listened for a while when Cousteau proclaimed, "Ze oceans are dying," but Cousteau is gone now, and there's no one to take his place, though President Obama on the morn of his second inauguration finally seemed ready to try.

Anyway, to build on my father's work required an exercise in pure science.

Pure science is rarely earthshaking. It mostly advances in tiny, painstakingly researched, time-consuming increments. But the building blocks do mount up, and the more painstaking the science is, the more permanent they are.

The Chaplin Project has not only built on my father's work, its conclusions, limited in scope as they may be, are ones he would have loved. There is hope for the reefs. Incredibly enough, fish communities remain as diverse as they were fifty years ago. This discovery is our unique contribution to the great jigsaw puzzle that is science. We have confirmed that the means of population regeneration are still there, at least for the time being, until the dead coral that the fish inhabit crumbles away.

The marine protected area in the Exumas that my father helped found in 1959 highlights one avenue of redemption for this "degradation debt." Thanks to the booming population of algae-eating parrotfish, coral inside the reserve is now four times healthier than it is outside. This has all happened since fishing was banned there fifteen years ago, just a blip on the evolutionary time frame, as was the Dairy Bull Reef recovery off Jamaica. We think of reefs as timeless structures like mountains, but they're not. They can be destroyed or revived during a president's term in office.

It's been a long, strange trip since Dominique's first galvanizing phone call, and I've weighed myself in the balance many times, often with a sinking feeling. Have I finally redeemed myself and shouldered the legacy? Would my father have been proud of what I've tried to do? Not long ago I got an email from Stan Waterman, my father's best friend and diving buddy who had inherited his beautiful shotguns instead of me:

> You're alive! I had lost contact with you and was delighted to hear word about your return to the old collecting expedition sites. In the course of working on an essay about our expedition to Cocos Keeling Atoll many years ago and sponsored by the Academy I contacted John Lundberg in

> order to reach Bill Smith-Vaniz. He was with us on that trip. John was most helpful in his reply and just happened to mention your own collecting work. Bravo! The spirit of Charlie is alive and well in your care.

Every second I've spent in and under the water for the Chaplin Project has been magical. I hear the symphony of the reef more clearly now than I ever did as a child, and I can detect new keys, chords, and progressions. The fish are numinous now, not just luminous: each one has a deeper meaning and its beauty signifies something beyond itself.

This is partly because each fish fits into our study: the scientifically proven fact that a species is still around after fifty years. But there's also the supernatural dimension recorded by Shakespeare in *The Tempest*. For me, it applies to two people I've loved who are now part of the element:

> Of their bones are coral made;
> Those are pearls that were their eyes;
> Nothing of them that doth fade,
> But doth suffer a sea-change
> Into something rich and strange.

My father and my partner Susan were with me in rich and strange ways every time I entered the water, especially in the most edgy, untamed place available: the Clifton Point drop-off.

They both loved the frisson of danger and excitement that comes from putting yourself outside the protections of civilization, knowing that you're no longer at the top of the food chain, that beneath you is an abyss into which you could disappear without a trace.

I love it too. And I've come to realize it's not the danger and excitement that really draws us, it's the feeling of being part of something much larger than ourselves. Thoreau said it best: "In wildness is the preservation of the world."

Reprise

November 2010

In science, the more replications you can make, the better the data. Four years after the expedition that led to her first paper, Katriina organized another. We'd go back to the same four sites off the north shore of New Providence Island, investigating now not only how things changed since the original Böhlke-Chaplin collections, but how they've changed since Loren Kellogg and his team made theirs.

The chaotic days of the Kellogg collections were ancient history. Katriina ran a tight ship, assisted by her parents Riivo (a chest surgeon) and Maia, Estonian immigrants who are living the American dream in Canada. Besides the veteran Heidi Hertler and myself, expedition members included the impressive Harrison Ford look-alike Mark Westneat, forty-eight, curator of fishes and director of the Biodiversity Synthesis Center at the Field Museum in Chicago, and the up-and-coming Ron Eytan, thirty-two, soon to be a postdoctoral researcher at Yale specializing in the origin and maintenance of marine biodiversity. It was a topflight team. In advance, Katriina had prepared two thousand prelabeled tissue vials. She intended to set a record.

Even with the tightest of ships, though, there are always problems in a scientific expedition. First, our precious rotenone was held up for a day and a half in customs. Then, after we liberated it with a morning-long expedition to the airport, a howling northeast wind kicked up and we had to scrub the next day. Ditto our third day: the ship channel into Nassau Harbor became impassable even for ocean liners. We collected fish at the fish market at Potter's Cay.

On our fourth day, the surf was bigger than I'd ever seen it, even as a kid. Huge ten-foot swells broke over Salt Cay as I watched from Cabbage Beach near Atlantis. The surf on the beach itself was overhead. Beach chairs that had been sucked into the waves made swimming dangerous even if one had wanted to brave it. I took refuge in a nearby restaurant, where I read that Sir Sol Kerzner (lately knighted) had been so piqued at the government's recent decision to grant a permit to a Chinese company to build a $400 million resort at Cable Beach that he threatened to cancel "Phase 4" construction at Atlantis. However, the story concluded, a rumor that Kerzner was planning to sell out completely was "unfounded."

On the fifth day the swell was still huge, but we were running out of time and decided to chance it. Breaking waves made the channel tricky, but Captain Roscoe expertly picked out a route and we made it with only one near swamping (a wave actually broke over the boat's flying bridge). Anchoring in thirty feet of water near the mid-depth Delaporte Point site, Captain Roscoe shook his head. "Goin' to be pretty roiled up down dere." He was right: when I SCUBAed down to reconnoiter, visibility was about six feet and the heavy surge gyrated me through uncomfortable ten-foot arcs over the bottom, smashing me into coral protrusions. "We're wasting our time here," I announced, back on the surface. Katriina agreed, but only after she went to see for herself. To this group I was only an amateur, not to be trusted with executive judgments or even to help put down rotenone. It rankled a bit, but I guess that's the price you pay for real professionalism.

We chugged fifteen miles east to the fifty-foot-deep Green Cay head, my father's resting place, praying that the surge there wouldn't be as extreme. Miraculously, it wasn't, though instead of hanging in the water column like astronauts we found ourselves on underwater trapezes, one minute swinging close in to the head, the next swinging gracefully away. Visibility was pretty good, around sixty feet, and the surge actually helped the collection: washing the emulsified rotenone into crevices where it wouldn't normally have penetrated, and washing the specimens out. Timing was everything: a specimen would appear, you'd have an instant to net it, then it would be sucked back into the reef and you'd be swung away into open water.

Heidi was in charge of laying transects to measure coral cover, not so easy when you're flying around on a trapeze. I could see laughter bubbles issuing from her regulator as she swung helplessly back and forth. My father would have gotten a chortle out of it too. We had heard stories about coral regeneration in the four years since our last trip, but, sadly, the coral looked in worse shape than before. Algae had gained important ground: now the spires were mostly covered, while four years ago they were among the last of the living coral. But . . . when we first entered the water, a large barracuda hung over the topmost spire. Was it the same one I'd seen in the cloud of my father's ashes?

Anyway, it seemed like a blessing. We made a good haul: 859 specimens, representing seventy-one species, only a little fewer than four years ago. We collected no saddle squirrelfish this time, but Ron Eytan thought we might have a new species of goby. Only one man in the world, Victor Springer, a contemporary of Jim Böhlke's, could say for sure. Katriina spotted a possible new cardinalfish.

Loren's team had done most of the specimen-handling aboard the *Guanahani* on the way back to port, but Katriina organized a set of trays on picnic tables in the pool area of our motel for sorting, photographing, sampling, labeling, and finally preserving. Feeling a little bent from too much time at depth, I crashed at ten p.m., but the rest of the team stayed up until three thirty in the morning. Specimen-handling was clearly their obsession, for no one more than Ron, the expert on tiny gobies and blennies. Wearing a headlamp, his long hair streaked with gray, looking deranged, he arranged hundreds of specimens no bigger than paper clips in perfectly ordered rows. "I love them," he said with a grin at one point. "They're where the action is."

The surge dropped to more manageable levels, and we had two more successful days of collecting at other sites. In the evenings, at the sorting tables, I felt challenged to stay up and help. I found there is a seductive rhythm to doing the same thing over and over, with intense attention to detail: a kind of physical mantra that feeds on itself and eventually puts you into a new dimension. Have I finally discovered my father's secret? How he could put in those long hours in the lab that made him a scientist? How scientists can keep operating in the field on only three hours of sleep a night? Maybe the zen of repetition is the answer—this, and the jokes and stories that you share with your co-conspirators. For example, Mark's tale about getting attacked by an angry male dugong after someone had speared its mate; Ron's about discovering a live scorpion in his face mask at depth.

In the end, Katriina's team sampled some 1,400 specimens from a total of 2,000 collected over three days, while Loren's sampled 255 specimens from a total of 5,428 collected over nine. But who's keeping score?

Katriina's paper on this trip was published in the March 2013 bicentennial edition of the *Proceedings of the Academy of Natural Sciences of Philadelphia*, volume 162. As a side issue, she surveyed Bahamian fish collections at the country's seven top institutions and found the academy's to be by far the largest: with 61,246 specimens, way more than the nearest competitor, the American Museum of Natural History in New York, with only 54,065. Most of the academy's huge collection was put together by my father, Jim Böhlke, Loren Kellogg, Katriina, and me.[12]

Generally, her findings confirmed and expanded trends outlined in her 2010 *Bulletin of Marine Science* paper—increased population percentages of herbivores like parrotfish, decreased percentages of planktivores like damselfish and cardinalfish—consistent with a degraded reef habitat. And sure enough, Heidi's coral surveys showed decreased cover since Jaap's survey four years earlier.

There was one particularly ominous development:

> Another result of particular interest is the apparent inverse trend seen in *holocentrid* (squirrelfish) and *apogonid* (cardinalfish) relative abundance. Both groups are primarily nocturnal feeders that use caves and crevices for shelter during the day. Luckhurst and Luckhurst (1978) showed that the much larger-bodied *hololcentrids* have longer residency times than *apogonids* within particular coral shelter. Thus, even though *apogonids* are not habitat specialists, a possible explanation for the observed trends is that the small *apogonids* are outcompeted for shelter space in degraded coral habitats. To our knowledge this inverse relationship between *holocentrid* and *apogonid* relative abundance has not been documented elsewhere.

The dead reefs are crumbling away, and the degradation debt is coming due as I write. As sheltering space contracts, the tiny cardinalfish may be getting squeezed out. According to this grim scenario, it's just a matter of time before the larger, more aggressive squirrelfish are squeezed out themselves. In the end, the collapsing reefs will be abandoned and the fish that lived in them will have no refuge. The world's most diverse ecosystem will have been destroyed.

Dramatis Personae

Chaplin family portrait by artist Fritz Janschka

If you can find an old 1968 edition of *Fishes of the Bahamas*, you are in for a treat: minutely detailed colorplates of some thirty-five species, painted by Viennese artist Fritz Janschka. (In the 1992 edition, half of these were lost and exist only in the index, causing a fruitless search through the otherwise faultless text.) The Janschka plates glow with vigor and life and make the fishes into indelible characters.

Janschka had been a founder of the Vienna School of Fantastic Realism. He came to the United States in 1949 as a young artist-in-residence at Bryn Mawr College on a fellowship from my uncle Cummins Catherwood's foundation (later discovered to have channeled CIA funds). I knew him through the bizarre illustrated letters he'd write to my parents and a dauntingly insightful Chaplin family portrait that my father had framed and hung in his bedroom. In the foreground of a Dalíesque landscape sits a young nude couple, the smiling man staring intently at the quizzical woman whose back is turned. The man is holding a dead songbird in a frying pan, the woman's right hand is holding a living songbird, and her left is covering the face of a baby girl lying on the ground. The girl is holding a snake in each hand; one of them is biting the woman's right nipple. A short distance away, a boy half hidden behind a twisted dead tree is shaking a heart-shaped rattle. Janschka is now Professor Emeritus of Fine Art and Fairbank Professor Emeritus of the Humanities at Bryn Mawr with works in the Philadelphia Museum of Art, the Fort Worth Museum, and museums in Zurich and Vienna.

It should be noted, though, that the lion's share of the artwork in *Fishes of the Bahamas* (black-and-white drawings of each specimen described, close-ups of particular features used to identify a species, and general body outlines for each family of fishes) was done by Steven P. Gigliotti, director of medical illustration in the Harrison Department of Surgical Research at the University of Pennsylvania's medical school. He spent twelve years completing 700 drawings for the book. I am happy to reprint his one colorplate, of the papillose blenny, *Acanthemblemaria chaplini*, which begins the series.

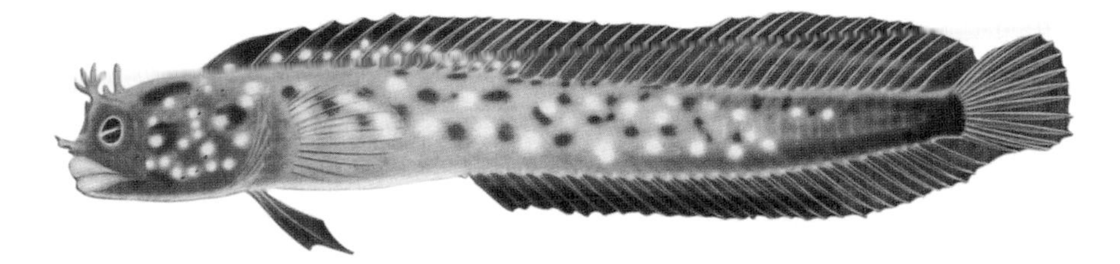

Frontispiece
Acanthemblemaria chaplini Böhlke
Papillose blenny

Illustrated specimen: 34.8 mm (1.4"). Grows to 1.8 inches.

Distinctions: This is the only Bahamian species in which elongate fleshy papillae replace the spines on top of the head and snout. Also, it has a higher number of anal-fin rays than the other forms (26 to 29 vs. 21 to 25). In general appearance, it is most like an *aspera*, but a glance at the top of the head will separate the two. The orbital cirrus of this form also is distinctive; in contrast to the arrangement of other species, where branches may arise from any or all sides of the cirrus, the branching in *chaplini* is most often in one plane and the branches are given off the main trunk as opposing pairs.

Coloration: Refer to the Frontispiece.

Remarks: This species frequents about the same depth range and general habitat as *maria* in the Bahamas; we have taken it at depths ranging from 5 to 38 feet. It inhabits limestone slopes rather than patch reefs; these slopes usually are dotted with small corals, sea urchins, etc.

Distribution: Known only from the Bahamas, where it has been taken on the Little Bahama, Great Bahama, and Cay Sal Banks.

Why they named this tiny, thick-lipped, snakelike, fleshy-antlered, faintly ridiculous fish after my mother is a question probably better left alone. I will point out, though, that Jim Böhlke gave *his* wife's name to the cleaning goby, *Gobiosoma genie*, a svelte black fish with glowing yellow longitudinal stripes and the endearing habit of picking parasites from the mouth, gills, and bodies of larger fishes, which have actually waited in line for the service.

Still, they gave the frontispiece of *Fishes of the Bahamas* to my mother's namesake (the original painting hangs in the Chaplin Library at the academy), as was only her due since she'd financed not only the volume's publication but also Jim Böhlke's salary and fifteen years of fieldwork.

I never saw her put on a face mask and look below the surface herself, though she was a good swimmer and actually rescued a young boy who'd gotten out of his depth off our beach. I never expected her to—it was as unthinkable as imagining a goggling Queen Elizabeth. But there must have been a better reason. I doubt she'd been frightened of the depths; before her crippling fall, she flew gliders, sailed yachts, skied untracked slopes, climbed mountains, and took her young son to witness at least one forest fire. Maybe she hadn't wanted to trespass on my father's territory? In every other respect, he was hers.

Not long ago, I took Sarah to see her grave in the cemetery at the Church of the Redeemer in Bryn Mawr, the first time I'd been back since her burial there in 1983.

It was a dark, drizzly November afternoon, and the lights of the cars passing on Old Gulph Road were already reflecting off its black asphalt. Sarah waited in the car: I wanted to be alone for the first few minutes of the reunion.

The drizzle was beading in my sweater and the light was fading as I walked under oaks and maples down a gentle slope past tombstones bearing the names of many of my mother's friends. In what I thought was her general vicinity, I cast about randomly with no success. I had never seen my mother's stone, and it was hard to be systematic—the graves were not arranged symmetrically, and, anyway, the scientific approach had never really worked in my personal life. I tried to search in parallel transects fifteen feet apart, keying them on landmark trees and larger gravestones, but kept getting distracted. Again and again I passed those family names—Roberts, Wood, Strazhupe, Montgomery, Scott, Morris. It was as if she were giving her final party, one that would last forever. And I was not invited.

I'm not sure when I started to cry, but it wasn't because I was soaked through and freezing. The whole episode seemed to fit perfectly into our heritage of failure to find and understand each other or to express our love. I searched unsuccessfully until it was too dark to see. Back in the car, Sarah put her arms around me and didn't say a word.

PLATE 9
Gramma loreto Poey
Fairy basslet

Illustrated specimen: 61 mm (2.4"). Grows to over 3 inches.

Distinctions: The coloration—anterior half violet, posterior half yellow—is not matched by any Bahamian bass; it is approached only by that of the young Spanish hogfish (*Bodianus rufus*), a wrasse. The threeline and fairy basslets usually have a somewhat emarginate caudal fin, while that of the black-cap basslet is deeply lunate and that of the spotfin basslet rounded.

Coloration: See Plate 9 for the color pattern of this spectacular little fish. It has been noted that in specimens from the Virgin Islands and Haiti the dark color of the head and shoulder regions extends farther aft than in Bahamian specimens.

Remarks: *Gramma loreto* is most commonly found in caves or under ledges. It swims with its belly toward the substratum, thus under ledges is seen upside down. If pursued, it usually will hide in a crack or small hole in the reef; frequently small groups of two or three to a dozen or more are seen in the same cave. It may be seen in water as shallow as a few feet where the water quality is pure; it has been collected or observed at depths to 110 feet in the Virgin Islands, to 190 feet in the Bahamas, and to 200 feet in Curaçao. It has been seen to move several inches away from the bottom and pick small items from the plankton. Eibl-Eibesfeldt reported *G. loreto* (as *G. hemichrysos*) among those fish he observed picking at the bodies of other fishes at Bonaire, allegedly in an attempt to feed on ectoparasites.

Distribution: Known from Bermuda and the Bahamas through the Greater and Lesser Antilles to the islands off Venezuela, and in the western Caribbean; apparently not present in Florida. In the Bahamas, it is common everywhere.

When my father showed me this little fish on my first dive at age six, he couldn't have known it was going to change my life. The grandiose idea of passing on a legacy was not my father's style.

To him, the fish was an *objet* to share with his son, all the more beautiful because it was alive, frequented dramatic settings, and had endearing habits. I've never seen them cleaning parasites from other fish, but they're very trusting after they get to know you. When you first peer into their caves, they dart into crevices leaving a fluorescent afterglow in your mind's eye, but if you can stay for a while (thanks to SCUBA) they'll reemerge and you can get within a foot or two of them. One actually came up and picked at the glass on my face mask with its delicate mouth. There are usually more than one in a given cavern, hanging there at different angles like living Christmas tree ornaments.

My father had an eye for *objets* (my mother's friends called him an *objet* himself, after all, and it takes one to know one), and I think it was the beauty of fishes that first attracted him to study them, more than a scientific motive. He delighted in anthropomorphizing them with such phrases as "lolling about" (French grunt), "slinking about the reefs in a very deliberate manner" (indigo hamlet), "fond of hiding among the blades of stinging coral" (yellowtail damselfish), "old friends" (any species common in the Bahamas), "old growler" (barracuda over five feet).

Anyway, the legacy was passed, not through daunting ceremony and solemn words but through a piece of living art. I saw what my father wanted me to see and took it from there.

Fifty years later, the fact that a descendent of that fairy basslet was not under that same ledge off the beach in front of the Chaplin House seemed a very grim omen to me, not to mention having almost been killed by a jet ski while searching for it. But the omen was not borne out. On the spectacular deep coral head near Green Cay, where I'd released my father's ashes, the relative abundance of specimens of the family *Grammatidae* (to which the fairy basslet belongs) in our collections was so close to its relative abundance in collections made by my father fifty years earlier as to be statistically the same. In all collections at that site, the proportion of fairy basslets to other species was second only to the cardinalfish.

My own eyes confirmed this. Cave after familiar cave in the dark underpinnings of the massive coral tower was lit with violet-and-yellow pinpoints of reflected light: my old friends were still there.

Totem: an animal, plant, or other object (adopted by an individual) serving as the emblem of a family or clan and often regarded as a reminder of its ancestry (Webster's).

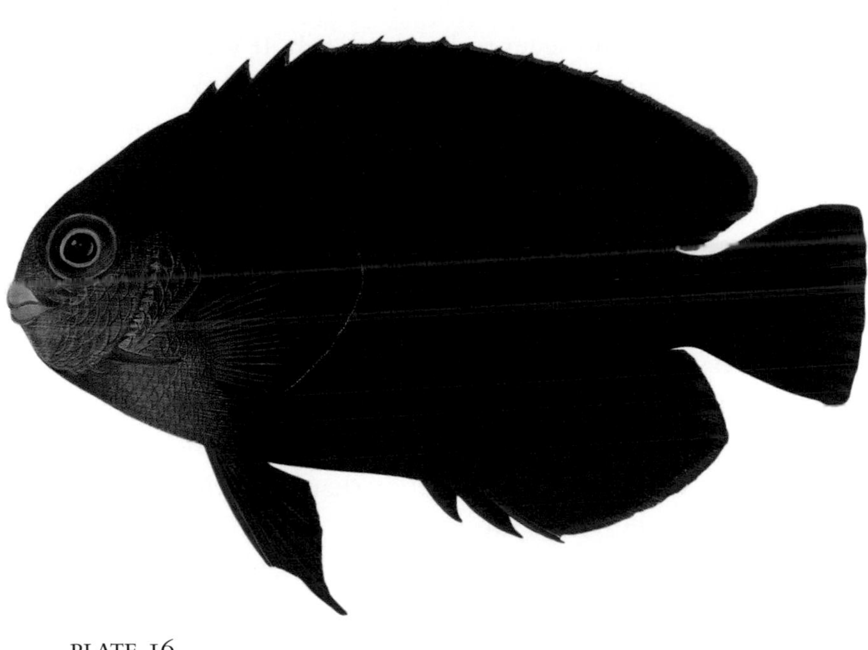

PLATE 16
Centropyge argi Woods and Kanazawa
Cherubfish

Illustrated specimen: 60 mm (2.4"). Grows to just over 2.5 inches.

Distinctions: The long spine on the preopercle marks this species as an angelfish, as opposed to a butterflyfish, and the strong preorbital spines, just below the eye, separate it from the other West Atlantic angelfishes. The color, as in all members of this family, is distinctive.

Coloration: (Plate 16) Very young specimens show a rounded dark marking just above the pectoral fin base, but on larger specimens this is largely obscured by the over-all darkening of the body. Otherwise, the young are similar to the adults, differing in this respect from other angelfish juveniles, which have an entirely different color pattern from the adults.

Remarks: This would seem to be a rare fish. It was first described in 1951 from a single Bermudian specimen taken in 1908. In 1952 another was retrieved from the stomach of a snapper off Yucatán, but no more were collected until 1959, when a number were netted alive near Bimini in the Bahamas by the Miami Seaquarium staff. Thus alerted to their presence in these islands, we did find a few more, among shallow-water conch piles in the fast-running waters of Nassau Harbor, and one each in deepwater poison collections off Grand Bahama Island and on Cay Sal Bank. There may be more than we think, for until 1959 we did not know they existed in the Bahamas, and might have confused them with the ever-present damselfishes, which they resemble in shape and habits. They are busy little fish, always on the move and hard to catch. They are attractive additions to a small-fish aquarium and appear to feed on algae.

Distribution: Known from Bermuda, the Bahamas, and off Yucatán. The Bahamian specimens are discussed above.

I was the one who first spotted this fish in Nassau Harbor, on the next to the last day of my Easter vacation in my first year at the boarding school to which I had been banished. My father and I were snorkeling on the wreck of an old cement barge in about fifteen feet of water west of our dock. Late afternoon: we were just goofing around before teatime. It was one of our favorite fish-watching spots; the barge had many overhangs and grottoes, rising about seven feet from the bottom, encrusted with plate and stinging corals. Conch fishermen had taken to dumping their empty shells around the wreck, forming additional shelter. Fish were all over the place in large variety: blue-and-gray chromis hanging over the high points, schools of French grunts over the conch shells, wrasse of all kinds gliding sinuously among the spires. You could often spot rarer fish like little harlequin bass, golden coneys, rock hinds, tobaccofish, and puddingwives. The water in the harbor was just as clear as anywhere else in those days, and there was always a current carrying plankton and possibly a touch of raw sewage from the Ryan (now Klenk) place on shore.

The little fish's movements were what first caught my eye: it had its own style of swimming, more of a flutter than other fish as it made its way busily among the conch shells in the deeper water near the wreck. Then I could see it had an unusual spade shape and a dark blue color. My heart started to beat faster. I wanted to dive down to get a closer look, but as soon as I jackknifed it noticed the movement and disappeared among the conch shells. I called to my father: "I think there's something different over here."

He swam over. "What did it look like?"

"A little blue-and-yellow fish. Kind of like a damselfish, but it didn't swim like one." I pointed to where I'd last seen it. "It was right down there."

The longer we waited, the dumber I felt. There were plenty of damselfish around: I'd probably made a mistake, a misidentification. But I'd been so sure. I would have bet my life.

I could see my father check his Rolex. Teatime! My mother would be sitting on the south verandah, ready to pour, getting more and more impatient. She'd invited the Klenks, too (this was before "Gold diggers together, hey Charlie?").

There! Ten feet away from where I'd originally seen it, the little dark blue fish was fluttering about as if it had always been there. I poked my father's shoulder and pointed. We watched without moving, even holding our breath (at least I was). Then he turned, nodded, gave me the circle of perfection sign, and we cautiously swam closer. Almost as soon as we started, the little fish disappeared again but not before I'd noticed the bright blue circle around its yellow eye, the orange mask over its nose and mouth, and the iridescent blue strip on it dorsal and anal fins.

My father removed his mouthpiece and raised his head above water to talk: "Well spotted, old fellow! Well done! A miraculous find."

"What is it?"

"Ten to one it's a centropyge. The first we've ever seen here. Cherubfish is the common name. They're an angelfish, not a damselfish at all."

"Do you think we could catch it?" I could already see the little creature making itself at home in one of our spacious new laboratory aquariums.

"Very difficult in those conchs. And it's wary as a fox, the little brute. We'll give it a shot, though. We might get it with a gill net if God is on our side."

"Now?"

He checked his watch again. "Lord no! We're late for tea already. Tomorrow, first thing."

"But . . . what if it's not here tomorrow?"

"It will be. Don't you worry."

Oddly enough, when my father made assurances like that, I believed him in spite of some evidence to the contrary. For example, he swore they'd still be making MG TCs when I was old enough to drive. He described the centropyge to the Klenks in glowing terms, giving me full credit for the discovery. "He's going to grow up to be a carbon copy of his father," my mother said with a typical note of disapproval. My sister was away at the time, I forget where.

Hope Klenk had a cute but bitchy blond daughter from her Ryan marriage, a little younger than me. "Alix would love to watch the capture, I'm sure," she said. "Do you think it would be all right if she tagged along?"

"Of course," my father said. "We'll leave at about ten from our dock."

"Not until ten?" I was outraged, and the idea of Alix getting in the way didn't help.

"We have a lot of gear to prepare, don't forget."

I spent a lot of the night wondering what centropyges do when it gets dark. Do they sleep in conch shells? Do they hole up in the coral? Do they burrow under the sand? Or do they just swim around in ever-increasing circles until they disappear? Why were they so rare, anyway? There had to be a very good reason.

We had an early breakfast and started assembling the gear we'd need: an aqualung and weights for my father, snorkeling gear and small free-diving weights for me (I pleaded for an aqualung myself, but my father said he wanted me more mobile), dip nets, chivy sticks, two collection buckets in case we got other fish we wanted to keep separate from the centropyge. And the gill net, a ten-foot-long, three-foot-high curtain of coarsely woven almost invisible nylon filaments with small lead weights on the bottom and buoyant pieces of Styrofoam on the top. The idea was to chivy the fish into it with a sudden movement and grab them before they could get free.

By the time we'd carried the gear from the laboratory to the dock and loaded it into our Boston Whaler, it was indeed ten o'clock. But Alix was nowhere in sight. "Why don't you run over and alert her, Gordy," my father said.

I ran as fast as I could along the concrete path that led past the high pink wall separating our properties. The Klenk bungalow was similar to ours, but fancier . . . done up like a Southampton beach house. Alix, in a pink silk, lace-collared peignoir, was alone at the breakfast table with a white-shelled boiled egg in a white eggcup in front of her. "It's ten o'clock," I panted. "We're ready to go." She'd just arrived from New York for her own Easter vacation; I hadn't seen her for a year or so.

"Mummy didn't wake me up." She made a little moue. "She and Uncle Cliff went to town early."

I just stood there, feeling my face get red.

"I'll be down in a minute," she announced. "After I finish breakfast and get into my swimsuit."

"What?"

"I said: I'll . . . be . . . down . . . in . . . a . . . minute. Okay?"

I couldn't think of a thing to say back, so I turned and left. About half an hour later (or, more precisely, at 10:39) we saw the glimmer of her green floral-patterned Lilly Pulitzer swimsuit appear from beneath the Klenk palm trees and round the pink wall. "*Duh-hunh.*" My father laughed his signature laugh. "You'll get used to waiting for women, my lad." As far as I knew, Alix was barely twelve years old.

I already had my snorkeling gear on, chivying stick and dip net in hand by the time we anchored near the wreck, and was over the side and into the water before my father had even shut down the engine. My job was to locate the fish. Then my father would submerge with his aqualung and the net.

Just as I'd feared, the centropyge was no longer in the same place. It had probably moved on. I swam around with a sinking heart, checking the dinghy every once in a while to see my father showing Alix how to fit her snorkel onto the headband of her face mask—she didn't even know how—as if nothing was wrong.

He'd be sor— No, Gorblimey, *there it was*, not over the conch shells this time but near a ledge in deeper water, maybe twenty feet down, doing its enchanting flutter dance as it went about its errands like a woman popping into shops.

By the time my father got there with the net, the fish had disappeared under the ledge. I dove down and pointed out the spot, and when I surfaced Alix flapped over with her yellow swim fins. "I didn't know you could dive like that." She giggled. "You looked like a fish yourself."

I didn't answer. Quite a respectable way below us, my father was arranging his net around the ledge: he'd move a little way off when finished until the fish gained confidence and reemerged. Then, coming at it from above and below, we'd try to startle it into the net.

Alix tried a shallow dive to about eight feet, the top of her bathing suit billowing interestingly as she came back up. "Don't *do* that," I called. "You'll scare the fish."

"They didn't look scared."

"That's because they're not the right fish, dumbo. The one we want is hiding under that ledge. It's really shy."

"I'm *not* dumbo. What does it look like?"

I thought for a minute. "Like a jewel. Something from Tiffany's. Only better."

It was now a waiting game. As we floated around, my attention was divided between the ledge and the top of Alix's bathing suit—a lot more revealing than she probably realized. What I could see inside it was brand-new and put her in an entirely different category of creatures from the one I was familiar with. Like the centropyge itself. And my heart pounded oddly for both of them.

At last I could see the little fish poke its head out, look warily around, hesitate, then flutter to the top of the ledge and begin to nibble at the algae there. It seemed oblivious to the net, only three feet or so away. Breath shortening, I tapped Alix on the shoulder and pointed. Our arms brushed lightly as she turned to look.

My father, seated on the bottom about ten feet away, slowly inclined toward the ledge until he was lying prone, inching his long chivy stick out in front of him with one hand to cut the fish off and using the tips of the other hand's fingers to walk himself along. In a series of erratic zigzags, the fish moved a little farther from the ledge toward the net. Too good to be true.

I could see my father flex his knees for a lunge and started breathing deep to store oxygen for my own dive. When his fins snapped, I was on my way.

It had worked! The centropyge was flapping in the net, but it wasn't tangled. My father missed with his long-bellied dip net and then it was free, circling confusedly for a second in mid-water. I had my shot, and made it good. And I imagined I could feel admiring eyes on me from twenty feet up.

Alix, still in her damp bathing suit, came back with us to install the centropyge in a large tank in the laboratory that we'd kept free of aggressive damselfish that might chivy it about and nibble on its fins. It was a tank for placid algae-eaters and featured large algae-covered rocks on a white sand bottom. The little fish looked completely at home among the wrasse, butterflyfish and juvenile angels, and was soon fluttering about nibbling at the algae itself. "You were right," Alix told me after my father had gone up to the house to get ready for lunch. "It does look like something from Tiffany's."

"But better. Right?"

"What could be better than something from Tiffany's?" She grinned and wrinkled her nose. "At least to a *girl*. But you caught it . . . That was really amazing. You're such a good diver!"

I blushed and shook my head. "Twenty feet isn't that deep."

She came closer and stared up into my face. "You're leaving tomorrow, aren't you? To go back to school? Why doesn't your stupid school have vacations the same time as the rest of us?"

"I don't know. It's horrible. I hate it."

"Do you know something weird? I think I'll miss you. Will you miss me?"

"Well . . . yeah!"

"So . . . here's something to remember me by." Suddenly she put her arms around my neck, pulled my head down, and gave me a dry, soft, schoolgirl kiss on the lips: my first one. Whenever I looked in the tank, I remembered it. Close up, you can see a centropyge's mouth is perfectly shaped for kissing. Maybe that's why they're known as cherubfish. That was the only centropyge I ever saw. I wasn't on the trips to Grand Bahama Island and Cay Sal Bank, where my father and Jim Böhlke collected two more on deepwater poison stations. But they collected none on the four New Providence Island stations we chose to replicate in our retrospective study, and we collected none fifty years later. Are they extinct now, because of pressure from aquarium collectors? It's almost impossible to say for sure: fish are much harder to keep track of than terrestrial creatures. But the question haunts me, along with the memory of that first kiss.

As for Alix, she married a banker and died a tragic death.

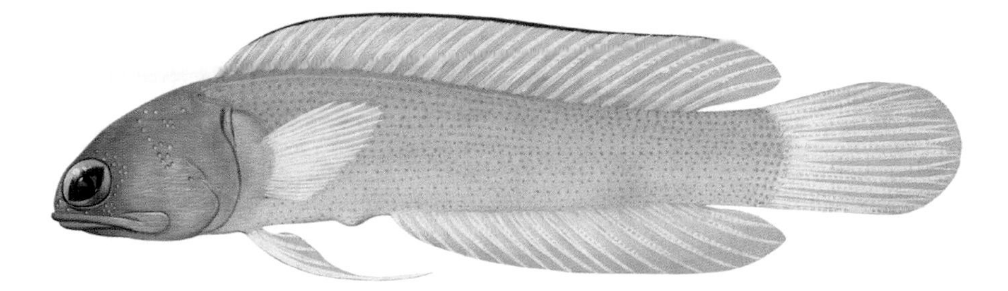

PLATE 31
Opistognathus aurifrons (Jordan and Thompson)
Yellowhead jawfish

Illustrated specimen: 70.3 mm (2.8"). Grows to about 3.7 inches.

Distinctions: Color pattern, fin shapes and behavior all readily identify this species among Bahamian jawfishes. The absence of dark marblings or other markings from the body and fins is unique in the islands to this species, except for the yellow females of *gilberti*. The soft dorsal, anal, caudal and especially ventral fins are elongated (more so as this fish increases in size), giving the fish a graceful appearance. Its habits are referred to below. The yellowhead jawfish has several greatly enlarged and recurved canines on the lower jaw; the others do not, except for *gilberti,* in which they are not as well developed.

Coloration: (Plate 31). The spotted effect shown on this plate is not noticed in casual observation of the smaller live or dead fish, which appear generally pale greyish blue. In young, the yellow color tends to be centralized in a small area on top of the head; the blue color of the body and fins is not yet developed in this size—about one inch.

Remarks: *O. aurifrons* has been taken at depths between eight and 115 feet in the Bahamas (to 135 feet in the Florida Keys), while the most common jawfishes in the islands, *whitehursti* and *maxillosus*, often are found in numbers at only two or three feet. The burrows of *whitehursti* and *maxillosus* most often are in areas of firm limestone substrate, while those of *aurifrons* are in crushed coral or sand and their walls can readily be pushed aside with the hand. *O. whitehursti* and *maxillosus* frequently inhabit the same beds, while *aurifrons* has colonies of its own farther down. *O. aurifrons* normally maintains itself six or eight inches above its burrow, tail down, revolving slowly, snapping at small organisms in the water; the other species stay within their burrows or sit in them with only the forepart of the head protruding. The male of this species broods the eggs orally; he deposits the egg mass in the burrow when he comes out to feed.

Distribution: Recorded from the Bahamas, Florida, Cuba, and the Virgin Islands. In the Bahamas, we have it from the Little Bahama, Great Bahama, and Cay Sal Banks, and from Hogsty Reef.

Stan Waterman, my father's best diving buddy and my childhood hero, lay on the sand bottom about twenty feet down near the lair of a jawfish. He'd been filming the three-inch-long, yellow-headed, big-mouthed creature for more than an hour and nothing it did was boring. It was grumpy, short-tempered, and snappish as it hung vertically above its carefully made burrow looking for food or driving away intruders. Every once in a while it would fussily swim down, pick up one of the retaining pebbles that formed the lip of the burrow, and move it a fraction of an inch this way or that. Sometimes it would dive headfirst into the burrow, dredge up a mouthful of sand, take it outside, and spit it onto the burrow's sloping banks. Finally, as the sun was going down, it emerged with its mouth stuffed with transparent eggs that were so close to hatching Stan could see the tiny larval fish inside each sac. As the light faded and Stan's air ran out, the jawfish began to spit out his babies, so small they looked like tiny swimming gnats.

"The miracle of life, no less," he announced to my father and me back on the surface. "And I couldn't get it on film." He raised a fist from the water and shook it. "Why couldn't the little *maladroit* have cooperated?" Here he parroted one of my father's favorite lines: "May the fire of Saint Anthony dart up his fundaments!"

It was a revelation to me that this heroic figure, who had actually speared a tiger shark with a Hawaiian sling, among other adventures, could be so fascinated with a three-inch-long creature that looked like Elmo on Sesame Street. "By God he was actually *growling* at me, Gordon." Stan's laughter came in great, windy gusts that carried everything before them. "I was *terrified*." He had a comedian's wide mouth that curled up at the corners connected by a long, straight, elastic upper lip and set in a bony, piratical face. His bald head was as streamlined as an underwater creature's, on top of a track star's body. I worshipped the ground he walked on, or better yet, the water he swam in.

We fell out of touch after Nassau. Stan's greatest adventure, filming *Blue Water, White Death*, a compulsive, bizarre, death-defying worldwide search for the great white shark, happened while I was working as a reporter in Vietnam in the late sixties, and I thought in a snobby way that what I was doing was more important. Truth to tell, I'd felt more than a little disenfranchised by my father's close friendship with Stan. But his agonized face, when I recognized my own initials on one of those shotgun cases at his farm, and his desperate offer to give me back the guns that should have been mine, allowed me to recoup by declining the offer and turned me into a lifelong friend in my own right. His accolade for the Chaplin Project—"The spirit of Charlie is alive and well in your care"—was the most important I could have had.

In the spring of 2012, I got an amazing invitation. Stan wanted me to join him, his friends, and family, to celebrate his ninetieth birthday diving with great white sharks off Guadalupe Island, Mexico. At this venerable age Stan was still actively leading dive trips around the world, though he had lost one eye to failed cataract surgery, his hearing was poor, many of his teeth were missing, and he was so stiff from bends-inflamed arthritis he needed a wheelchair to get through airports.

While filming *Blue Water, White Death* in the late sixties, the expedition leader Peter Gimbel had left the cage without warning to film a feeding frenzy of sharks around a dead whale. Stan had gone with him. Miraculously, they both survived, but the sharks were not great whites: "I wasn't trying to prove a thing," Stan said, countering one accusation of silly risk-taking. "Sometimes there's a job to do, that's all, and I felt my job was to stay with Peter." On this trip, assuming that the now-endangered shark showed up at all, I could imagine quite clearly the door to our cage swinging open and Stan taking his last chance to swim freely with his totem. It would make a great story, the kind told with relish by my father, and the story line would call for me to keep him company in my role as stand-in. Unless he made it clear he'd rather be alone.

As it turned out, I didn't have to worry. Stan's cage mate was the famous Australian diver Valerie Taylor, who with her husband, Ron, had helped film *Blue Water, White Death*. Ron had died a few months earlier, and Valerie had flown all the way from Sydney for Stan's birthday with a companion who was arranging travel trips around the moon. Valerie, now in her late seventies, an underwater Vanessa Redgrave with square chin and regal bearing, was not happy to be "stuck in a stupid cage" and very likely Stan wasn't either, though he was too polite to say so. But they had no choice: leaving it was against the law in this protected area, and anyway they were tethered to the cage with a hookah rig instead of free-swimming SCUBA.

I was probably the only guest aboard the dive boat who had never seen a great white shark, and when its awful smile slowly took shape through the misty blue like the Cheshire Cat in reverse I understood what all the excitement was about: Stan's totem was the scariest thing I'd ever seen, something out of a nightmare. *Implacable* is the common cliché, but it fits: the great shark slowly circles the cages all day, never gives up. It's like fate. You don't tempt it, at least I don't, though after they surfaced Stan's and Valerie's eyes were sparkling with an oddly childlike excitement. As they hugged each other on the fantail of the dive boat, they both looked as if they'd drunk from the Great White Fountain of Youth.

As for me, watching this awesome creature from the safety of my cage, I couldn't help thinking of Aldo Leopold's description of the jaguar that used to frequent the delta of the Colorado River in the twenties:

> His personality pervaded the wilderness; no living beast forgot his potential presence, for the price of unwariness was death. No deer rounded a bush, or stopped to nibble pods under a mesquite tree, without a premonitory sniff for *el tigre*. No campfire died without talk of him. No dog curled up for the night, save at his master's feet; he needed no telling that the king of cats still ruled the night; that those massive paws could fell an ox, those jaws shear off bones like a guillotine.

Paradoxically, *Blue Water, White Death* put the shark on the Endangered Species list, where it still is, along with the jaguar. Stan's friend Peter Benchley was inspired to write *Jaws* after watching the spectacular shark footage, Steven Spielberg weighed in, and the rest is history: a feeding frenzy in reverse. Now there are only 130 great whites in the Guadalupe Island population.

Maybe it's inevitable that both Stan and I, reunited by the sea, are now concerned above all else with trying to save what's left. The Age of Extinction is weighing more and more heavily: the coral reefs we remember from the old days are now mostly dead, fish populations are way down, and the big fish that so used to excite us are glimpsed through a glass darkly. Population figures on the great white are just starting to emerge, and the first estimates, published in 2011, are pathetically low. One population, located in the sizeable area around San Francisco Bay, number only 219 according to a census taken by photographing the sharks' distinctive dorsal fins. "The low number was a real surprise," said Taylor Chapple, the study's lead author and a doctoral student at the University of California, Davis, where the work was done. "It's lower than expected, and also substantially smaller than the populations of other large predators such as killer whales and polar bears." A genetically produced estimate of Australian great whites put that population at only 1,500, barely enough to be viable and less than half as many as previously thought.

Stan himself, and most of his famous and accomplished birthday guests, belong to an organization called Shark Savers, chaired by Peter Benchley's widow, Wendy. Benchley, before he died of pulmonary fibrosis, had turned into a fanatic conservationist.

PLATE 10
Gramma melacara Böhlke and Randall
Blackcap basslet

Illustrated specimen: 97.5 mm (3.8"). Grows to about 4 inches.

Distinctions: The black cap on the head, the black anterior dorsal fin, and the otherwise largely magenta body set off this species from other Bahaman grammids—in fact from all other Bahaman shore fishes. The blackcap basslet has the most deeply lunate caudal fin among Bahaman grammids.

Coloration: (Plate 10). The color pattern of the fish is essentially magenta or heliotrope with various black (a black with a magenta cast) markings, a few gold lines, and gold dots on the body. A black dorsal cap runs from the tip of the lower jaw back over the upper part of the head and onto the dorsal fins—broadest through the anterior portion of the spinous dorsal, decreasing in width distally till it forms a dark submarginal band on the fin which extends to the tip of the soft dorsal lobe.

Remarks: This is a deep-water species and often is exceedingly abundant where it occurs. It has been taken as shallow as 40 feet and as deep as 200 feet, being most common below 100 feet. The species is most common at the edge of the drop-off, the extremity of the bank. While blackcap basslets may occur in concentrations of a number of fish, they usually react as individuals rather than as a school; individuals typically stay close to the bottom and dart into small holes when disturbed.

Distribution: Known from the Bahamas and the islands off British Honduras. In the Bahamas, we have it from the Little Bahama and Great Bahama Banks.

The edge of the drop-off, where the blackcap basslet prefers to live, is a dividing line between two worlds. It's a place of choices. I feel closer to Susan there than anywhere else. If I choose to swim down over the edge and keep going into the darkening blue, we'd be together. But also I think she'd love it here on the edge. She was fascinated by edges.

The blackcap basslet is first cousin to the fairy basslet, the fish that changed my life. The shape's pretty much the same, but the blackcap's coloring is more subtle and in some ways more dramatic. Because of where it lives, its background tends to be the brilliant purple of open ocean, setting off its dark, luminous magenta in a way its shallow-water cousin can't match. Unlike its cousin's, there's something supernatural about its deeply forked tail. Jack Randall and Jim Böhlke, stretching the envelope of fish collecting in the fifties with the use of SCUBA gear, were the first to describe it.

Susan was a great one for looking rapt. Her mouth would open a little and the pupils of her eyes would expand like someone on acid. If we had ever dived together (SCUBA was too claustrophobic for her), that's the way she would have looked on the edge of the drop-off in a cloud of blackcap basslets.

For twelve years, after I left my wife and daughters for her, Susan and I lived in a rapt world. Kerouac himself would have been proud of us. It was indeed a wild, exhilarating ride . . . maybe too wild to last. Her favorite poem was by Emily Dickinson:

Wild Nights—Wild Nights!
Were I with thee
Wild Nights should be
Our luxury!

Futile—the Winds—
To a Heart in port—
Done with the Compass—
Done with the Chart!

Rowing in Eden—
Ah, the Sea!
Might I but moor—Tonight—
In Thee!

On Thanksgiving Day, 1992, I rang the fateful doorbell of a Beacon Hill town house. Susan's mother answered it and swept me into her arms before I could say anything. Then she took me into her ground floor apartment, and sat me down on the couch. I waited for her to ask me how her daughter had died, but instead she sat down beside me, put her hand on my arm, and said:

"So how are you, Gordon? You've certainly had a hell of a time, haven't you?"

I'd cried for Susan after she was torn from my arms in the typhoon, and on the next day as I searched for her body on the barren little island where I'd washed ashore. I'd cried for her on the day after that, when I made it back to the village on the other side of the lagoon where we'd been anchored before the typhoon struck. And I cried for her a few days later when we gave up the search and

I caught a relief vessel out to Kwajalein. I cried for her almost every day. But, sitting next to Susan's mother on her couch, feeling her hand on my arm, I couldn't help crying for myself. Her sympathy was so unexpected and amazing, I felt like I actually deserved it.

Susan's brothers and sister and many of her ten nieces and nephews have the long, narrow Atkinson nose, the thin straight lips, the small mouth, the deep-set gray eyes. Seeing them gathered in the living room upstairs on that Thanksgiving Day was like looking at an artist's variation on her own face.

I'd brought with me a computer printout of the typhoon's path that sympathetic weatherpeople on Kwajalein had made available, clearly showing the storm's unreported change of course that took the eye directly over our anchorage. We never would have stayed aboard that night if we'd known we were targeted. Our weatherfax reported the course about fifty miles north and hadn't registered the last-minute course shift.

It was like evidence in a trial for a case that had already been lost. After I'd passed the printout around and was answering her brother's questions, I imagined I could see the family slowly begin to experience Susan's absence as something horribly permanent instead of just a voice with bad news at the other end of a long-distance phone line. They stayed in touch with me for a short while after Susan's mother died of cancer, and the rest has been silence. I'd failed to save their beloved. I might even have been responsible for her death.

At her memorial service in Trinity Church, Copley Square, Susan's first cousin and his wife (who was also Susan's ex-sister-in-law) recited together a slightly altered version of "Ariel's Song" from *The Tempest*:

Full fathom five our Susan lies;
Of her bones are coral made;
Those are pearls that were her eyes;
Nothing of her that doth fade,

But doth suffer a sea-change
Into something rich and strange . . .

A depth of five fathoms would bring you close to the drop-off at Clifton Point, where the current generation of blackcap basslets plays in that same coral. It's deep enough to be still healthy.

PLATE 33
Gobiosoma multifasciatum Steindachner
Greenband goby

Illustrated specimen: 43.2 mm (1.7"), male, about maximum size for the species.

Distinctions: This little goby with green bands and a bright red stripe behind the eye is one of the most easily recognized species in the tropical West Atlantic. It is the only naked Bahamian species of *Gobiosoma* that is banded rather than striped. It has a filamentous first dorsal spine, except when young.

Coloration: See Plate 33.

Remarks: This is a fairly common species on limestone bottom in the Bahamas. Many of our collections are from tide pools in the limestone at shoreline. In Puerto Rico, it has been taken from dead branches of coral, and has been reported to hide among the spines of rock-boring sea urchins. In Curaçao it is said to live in sponges. No station at which we took this species was deeper than twelve feet and most were much shallower.

Distribution: Known from the Bahamas and Cuba south through the Lesser Antilles. In the Bahamas, we have it from the Little Bahama and Cay Sal Banks.

That's not entirely accurate. We also have it from the Great Bahama Bank, from a certain tide pool at Delaporte Point, about half an hour from the Chaplin House by boat. My sister, Susie, age four, was lucky enough to collect it. She'd turned over a little rock, and the panicked goby had blundered into her dip net by accident.

At that point, before we'd started to dive beneath the surface and were still wading in the shallows, the little goby was the most colorful sea creature we'd ever seen, like something out of a fairy tale. Our father carefully put it in a collecting bucket where it darted around in a streak of brilliant green. I myself was green with envy. He favored my sister with a grin and a pat on the back. "It'll be a great addition to the tank."

Even our mother took part in the ceremony of introducing the little goby to the world of our living room aquarium. After being dropped in, it made for the white sand bottom, where it perched alertly on the disk formed by its two fused ventral fins, checking out its new environment through its Batman mask of two red stripes that started at its nose, covered its eyes, and stretched back to the gill covers. It seemed content just to sit there, not swimming about like the other fishes. Only its gills and eyes moved. It had thick lips around a down-turned know-it-all mouth and a series of double chins; you expected it to make a weighty pronouncement or two.

Our mother laughed: "Why, he looks just like Michael Pakenham. Let's call him Michael."

My sister let out a howl. "No! It's a *girl*. Her name is *Alice*."

Our father raised his eyebrows. "Why Alice?"

"Alice in *Wonderland*."

Our mother turned and left the room.

When she chose to, Alice moved around the floor of the tank like a lizard, in little jumps. Winston the octopus was still in residence at the time, and Alice would jump unconcernedly past his conch shell while he stared into the middle distance with his glacial eyes, siphon expanding and contracting. He didn't want to eat Alice, that was obvious, though I wouldn't have cared if he had. I might even have enjoyed it. But Alice led a charmed life and was very voracious herself, pouncing on tiny morsels of conch that my father dropped near her.

By dinnertime, our mother was in a better mood. Okra was her favorite vegetable, and the lean, tangy Bahamian chicken and baby potatoes were crisp and brown. My father plucked the bottle of white Burgundy out of its cooler next to him, walked to her end of the table, and poured her glass half full with a flourish. "Montrachet, '47. Lightbourne's last bottle." He poured himself a full glass and raised it in a toast. "Here's how."

My mother sipped and smiled. "It *is* good. How clever of you to snap it up."

"Well, I was just a step ahead of the Lily Maid of Astolat."

"Is she the one that always wears a leotard so she doesn't get any sun?" my sister asked.

"Yes. Positively fish-belly white."

"I'd *hate* to be like that."

"Don't worry. You won't be," my mother told her. "You're as brown as a little native girl."

"Good."

"If you don't use lotion, when you get to be my age you're going to be wrinkled as a prune."

"Or a *raisin*," I said.

"But Gordon doesn't use lotion."

"He's a boy," my mother said. "It doesn't matter if he has wrinkled skin. Wrinkled skin looks good on men."

"Yeah," I said, checking out my father.

"I wish I was a boy." My sister stared at me.

I stared back. I could see her point.

"I *bet* you do," my mother said, making the word *bet* sound like *hate*.

She let the conversation drop while she attacked the food. Nobody could eat as fast as she could when it was something she liked—"gutsy" was the word my father used. Sometimes she'd rip out a fart, and he'd pretend to be shocked while she laughed. Years later, the eating scene in the movie *Tom Jones* would remind me of all this: very eighteenth century.

But speed was all to the good, as far as I was concerned. I was facing the corner window of the dining bungalow, and the setting sun lit up the tops of the palm trees along the walk. There was about half an hour of light left, and with any luck I could make it down to the dock in time to see the sun slide into the water at the west end of the harbor. I'd heard about the green flash but had never seen it. One of these days I would . . . unless I missed it.

My sister and I skipped the fresh pineapple dessert and ran to the dock as fast as we could to find the sun still a finger above the horizon. The water in front of it was red instead of blue, and the harbor boat traffic seemed annealed in a brand-new element.

"I think I saw it," my sister said when everything had turned violet.

"The green flash? No you didn't."

Fifty years later, I find myself on a twelve-hour flight from New York to the Dutch Antilles island of Saba, where my sister plans her triumphant finish.

We haven't talked in fifteen years; I've followed her career through family gossip and occasional reports in the press. One of them, in the *Washington Post* and titled "Solitary Susan," described how she'd pared her life down to the bare essentials: a small apartment, a bike, and a custom-built fifteen-foot paddleboard on which she spent a good deal of her time.

I used to find my sister's intense competitiveness mystifying, sometimes enraging. Why did she have to make even a friendly jog into a race? Or a session of sit-ups into a contest? Or a surfing outing into bare-knuckled jockeying for position? At one point we had a disastrous series of meetings with a counselor that uncovered a basic disagreement and left it on the table between us: I maintained our mother's cruelty came in the form of withholding love . . . and I got most of what little there was. My sister said it was physical. Physical cruelty. She wouldn't explain exactly what form this took, and that's when we stopped talking.

Since our childhood she'd gradually honed herself to a competitive edge that I couldn't hope to match. She and another woman (three months pregnant) made the first all-female ascent of the north face of the Grand Teton in Wyoming. She twice competed in the Hawaiian Ironman Triathlon (a consecutive 2.5 miles of ocean swimming, 100 miles of cycling, 26.5 miles of running—the land events through hot, black lava fields), placing respectably in her age group. Finally she took up long-distance paddleboarding, placing second in Hawaii's thirty-mile Molokai Channel race and third in the Santa Catalina–LA race.

I'll have to admit I thought her obsessions were getting the better of her when I learned of her quest to paddleboard the entire Windward and Leeward chain of islands from Grenada to Puerto Rico, some twenty-five interisland passages totaling over 400 miles, a project that would take years. Unquestionably she would be the first person in the world to accomplish this, but who cared?

I did, as it turned out.

I spend my flight time to Saba wondering and worrying. After fifteen years of silence, I have no idea who I'll find. I have a suspicion it will be someone so far out on the fringes of eccentricity that competition will no longer be an issue. Anyway, in the past year or so I've been feeling my age. I'm probably out of whatever strange race my sister and I've been running. But still, I fear that none of this will stop us from quickly reaching critical mass if we get into one of our . . . things.

Next morning, from the escort skiff in the deep blue of open ocean about five miles offshore, my binoculars pick out a tiny speck that materializes into a bright yellow fifteen-foot paddleboard supporting a red Lycra–suited figure with a Lawrence of Arabia headdress. My sister sits up, waves, and calls: "Hey, Gordon. Is that you?" Her voice hasn't changed at all.

We follow her in. She has a dogged paddling rhythm: twenty strokes prone, then twenty strokes kneeling. She explains later that this rhythm, anticipating the change, is what keeps her going. When she's prone, she rests her chin on one of three special pads made from foam rubber covered with plumbing glove latex. Choosing the pad adds crucial variety to the moment. Her headgear is designed for salt-water flyfisherman who need maximum sun protection. She wears bicycling tights, a Lycra top, and diving gloves to protect her hands from barracudas and stinging jellyfish. Only her face and her feet are bare. Strapped to her board are a GPS unit, a VHF radio, and a drinking bladder of water. She's been paddling for about six hours from Saint Eustatius. In the last couple of weeks, she's paddled from Montserrat to Nevis, Nevis to Saint Kitts, Saint Kitts to Sint Eustatius, some fifty-five miles.

The old concrete launching ramp in Saba's harbor is slippery and treacherous with sea moss. In my nylon shorts and tee shirt, I bottom-down it into waist-deep water and stand there as she rounds the breakwater and paddles toward me. She dismounts. When her feet are unquestionably on Saba soil, we put our arms carefully around each other. Her back feels hard as rock. I've brought some champagne, shake the bottle, and pour the bubbly over her head. We look at each other shyly and then look away.

Over the next few days, on the strange little island of Saba, I begin to see how she's used her oversize quota of family angst to become a certifiable character: people *do* care about her quixotic odyssey, strange as it is; she has friends and admirers all up and down the islands.

Physically, she's made herself into a female version of our father. I remember her struggling when we were younger to stamp out all vestiges of the femininity our mother had tried to encourage. Now she's on good terms with it: she looks like an extremely fit, muscular, mid-fiftyish Vanessa Redgrave.

She shed her husband (father of her son, Eli, who's accompanied her as support on much of her odyssey) a long time ago; now I have no idea what her love life is, and I don't feel like asking. It really doesn't matter. She writes articles for the prestigious *Surfer's Journal* and other outdoor publications, gives lectures, and generally promotes herself as the eccentric, uncompromising, slightly awe-inspiring character she's become. She's paddled her way out of the doldrums to confidence, strength, and clarity.

I envy that clarity. I do. And in spite of myself, I feel a competitive tug at my heartstrings. At the Saba airport, waiting a bit uncomfortably but lovingly to say our good-byes, I spot a copy of the Dutch Antilles *Daily Herald*. A reporter had covered my sister's triumphant landing, and we leaf through it eagerly to find the story. The lead paragraph reads: "*Paddler Susan Chaplin (64) was met by her older brother Gordon on Saba after her arrival from Statia Thursday at 1 p.m.*"

We stare at each other for a second and then at the same time burst into laughter. It's still there.

Endnotes

1. J. B. C. Jackson, "What Was Natural in the Coastal Oceans?" *Proceedings of the National Academy of Sciences* 98, no.10 (2001): 5411–18.
2. K. L. Ilves, L. L. Kellogg, A. M. Quattrini, G. W. Chaplin, H. Hertler, and J. G. Lundberg, "Assessing 50-year Change in Bahamian Reef Fish Assemblages: Evidence for Community Response to Recent Disturbance?" *Bulletin of Marine Science* 87, no. 3 (2011): 567–88, doi:10.5343/bms.2010.1052.
3. W. C. Jaap, J. M. Dupont, L. L. Kellogg, G. W. Chaplin, and H. Hertler, "Coral Reef Habitat Around New Providence Island, Bahamas," *Proceedings of the 11th International Coral Reef Symposium, Fort Lauderdale, Florida*, session XVIII (2008): 7–11.
4. G. K. Ostrander, K. M. Armstrong, E. T. Knobbe, D. Gerace, and E. P. Scully, "Rapid Transition in the Structure of a Coral Reef Community: The Effects of Coral Bleaching and Physical Disturbance," *Proceedings of the National Academy of Sciences* (USA) 97, no. 10 (2000): 5297–302, doi:10.1073/pnas.090104897.
5. K. L. Ilves, et al, "Assessing 50-Year Change in Bahamian Reef Fish Assemblages," 567–588.
6. L. Alvarez-Filip, N. K. Dulvy, J. A. Gill, I. M. Côté, and A. R. Watkinson, "Flattening of Caribbean Coral Reefs: Region-wide Declines in Architectural Complexity," *Proceedings of the Royal Society* 276 (June 2009), 3019–25, doi:10.1098/rspb.2009.0339.
7. J. A. Idjadi, S. C. Lee, J. F. Bruno, W. F. Precht, L. Allen-Requa, and P. J. Edmunds, "Rapid Phase-shift Reversal on a Jamaican Coral Reef," *Coral Reefs* 25 (March 2006): 209–11, doi:10.1007/s00338-006-0088-7.
8. P. J. Mumby, C. P. Dahlgren, A. R. Harborne, C. V. Kappel, F. Micheli, D. R. Brumbaugh, K. E. Holmes, J. M. Mendes, K. Broad, J. N. Sanchirico, K. Buch, S. Box, R. W. Stoffle, and A. B. Gill, "Fishing, Trophic Cascades, and the Process of Grazing on Coral Reefs," *Science* 311, no. 5757 (2006): 98–101, doi:10.1126/science.1121129.
9. T. P. Hughes, A. H. Baird, D. R. Bellwood, M. Card, S. R. Connolly, C. Folke, R. Grosberg, O. Hoegh-Guldberg, J. B. C. Jackson, J. Kleypas, J. M. Lough, P. Marshall, M. Nyström, S. R. Palumbi, J. M. Pandolfi, B. Rosen, and J. Roughgarden,

"Climate Change, Human Impacts, and the Resilience of Coral Reefs," *Science* 301, no. 5635 (2003): 929–33, doi:10.1126/science.1085046.t

10. J. B. C. Jackson, "Reefs since Columbus," *Coral Reefs* 16 (1997): supplement S23–S32.
11. P. A. Kramer, P. R. Kramer, and R. N. Ginsburg, "Assessment of the Andros Island Reef System, Bahamas (Part 1: Stony Corals and Algae)" *Atoll Research Bulletin* 496 (2003): 76–99.
12. K. L. Ilves, A. M. Quattrini, M. W. Westneat, R. I. Eytan, G. W. Chaplin, H. Hertler, and J. G. Lundberg, "Detection of Shifts in Coral Reef Fish Assemblage Structure over 50 Years at Reefs of New Providence Island, the Bahamas, Highlight the Value of the Academy of Natural Sciences' Collections in a Changing World," *Proceedings of the Academy of Natural Sciences of Philadelphia*, 162, no. 1 (2013): 61-87, doi:10.1635/053.162.0105